Lecture Notes in Computer Science 14790

Founding Editors

Gerhard Goos
Juris Hartmanis

Editorial Board Members

Elisa Bertino, *Purdue University, West Lafayette, IN, USA*
Wen Gao, *Peking University, Beijing, China*
Bernhard Steffen , *TU Dortmund University, Dortmund, Germany*
Moti Yung , *Columbia University, New York, NY, USA*

The series Lecture Notes in Computer Science (LNCS), including its subseries Lecture Notes in Artificial Intelligence (LNAI) and Lecture Notes in Bioinformatics (LNBI), has established itself as a medium for the publication of new developments in computer science and information technology research, teaching, and education.

LNCS enjoys close cooperation with the computer science R & D community, the series counts many renowned academics among its volume editors and paper authors, and collaborates with prestigious societies. Its mission is to serve this international community by providing an invaluable service, mainly focused on the publication of conference and workshop proceedings and postproceedings. LNCS commenced publication in 1973.

Abdelkader Hameurlain · A Min Tjoa ·
Reza Akbarinia · Angela Bonifati
Editors

Transactions on Large-Scale Data- and Knowledge-Centered Systems LVI

Special Issue on Data Management - Principles, Technologies, and Applications

Springer

Editors-in-Chief
Abdelkader Hameurlain
IRIT, Paul Sabatier University
Toulouse, France

A Min Tjoa
Technical University of Vienna
Vienna, Austria

Guest Editors
Reza Akbarinia
Inria and LIRMM
Montpellier, France

Angela Bonifati
Claude Bernard Lyon 1 University
and Institut Universitaire de France
Villeurbanne, France

ISSN 0302-9743　　　　　ISSN 1611-3349 (electronic)
Lecture Notes in Computer Science
ISSN 1869-1994　　　　　ISSN 2510-4942 (electronic)
Transactions on Large-Scale Data- and Knowledge-Centered Systems
ISBN 978-3-662-69602-6　　　ISBN 978-3-662-69603-3 (eBook)
https://doi.org/10.1007/978-3-662-69603-3

© The Editor(s) (if applicable) and The Author(s), under exclusive license
to Springer-Verlag GmbH, DE, part of Springer Nature 2024

This work is subject to copyright. All rights are solely and exclusively licensed by the Publisher, whether the whole or part of the material is concerned, specifically the rights of translation, reprinting, reuse of illustrations, recitation, broadcasting, reproduction on microfilms or in any other physical way, and transmission or information storage and retrieval, electronic adaptation, computer software, or by similar or dissimilar methodology now known or hereafter developed.
The use of general descriptive names, registered names, trademarks, service marks, etc. in this publication does not imply, even in the absence of a specific statement, that such names are exempt from the relevant protective laws and regulations and therefore free for general use.
The publisher, the authors and the editors are safe to assume that the advice and information in this book are believed to be true and accurate at the date of publication. Neither the publisher nor the authors or the editors give a warranty, expressed or implied, with respect to the material contained herein or for any errors or omissions that may have been made. The publisher remains neutral with regard to jurisdictional claims in published maps and institutional affiliations.

This Springer imprint is published by the registered company Springer-Verlag GmbH, DE, part of Springer Nature
The registered company address is: Heidelberger Platz 3, 14197 Berlin, Germany

If disposing of this product, please recycle the paper.

Preface

This volume contains a selection of fully revised papers presented at the 39th Conference on Data Management – Principles, Technologies and Applications (BDA 2023). For this special issue, we selected five articles covering a wide range of timely data management research topics on adaptive learning, personal data management systems, topic discovery in large corpora, spatio-temporal query processing, and data generation.

All authors were invited to prepare and submit a journal version of their contributions which were then fully re-reviewed by the editorial board of this special issue.

We would like to take this opportunity to express our sincere thanks to all authors and the editorial board of this special issue for their effort and their valuable contribution in raising the quality of the camera-ready version of the papers.

Finally, we are grateful to the Editors-in-Chief, Abdelkader Hameurlain and A Min Tjoa, for giving us the opportunity to publish this special issue as part of the TLDKS journal series.

May 2024

Reza Akbarinia
Angela Bonifati

Organization

Editors-in-Chief

Abdelkader Hameurlain Paul Sabatier University, IRIT, France
A Min Tjoa Technical University of Vienna, IFS, Austria

Guest Editors

Reza Akbarinia Inria & LIRMM, France
Angela Bonifati Claude Bernard Lyon 1 University & Institut Universitaire de France, France

Editorial Board

Reza Akbarinia Inria & LIRMM, France
Dagmar Auer Johannes Kepler University Linz, Austria
Djamal Benslimane University Lyon 1, France
Stéphane Bressan National University of Singapore, Singapore
Mirel Cosulschi University of Craiova, Romania
Johann Eder Alpen Adria University of Klagenfurt, Austria
Anna Formica National Research Council in Rome, Italy
Shahram Ghandeharizadeh University of Southern California, USA
Anastasios Gounaris Aristotle University of Thessaloniki, Greece
Sergio Ilarri University of Zaragoza, Spain
Petar Jovanovic Universitat Politècnica de Catalunya, BarcelonaTech, Spain
Aida Kamišalić Latifić University of Maribor, Slovenia
Dieter Kranzlmüller Ludwig-Maximilians-Universität München, Germany
Philippe Lamarre INSA Lyon, France
Lenka Lhotská Technical University of Prague, Czech Republic
Vladimir Marik Technical University of Prague, Czech Republic
Jorge Martinez Gil Software Competence Center Hagenberg, Austria
Riad Mokadem IRIT, Paul Sabatier University, France
Franck Morvan Paul Sabatier University, IRIT, France
Torben Bach Pedersen Aalborg University, Denmark

Günther Pernul University of Regensburg, Germany
Viera Rozinajova Kempelen Institute of Intelligent Technologies, Slovakia
Soror Sahri LIPADE, Descartes Paris University, France
Blerina Spahiu University of Milano-Bicocca, Italy
Joseph Vella University of Malta, Malta
Shaoyi Yin Paul Sabatier University, IRIT, France
Feng "George" Yu Youngstown State University, USA

Editorial Board of Special Issue

Oana Balalau Inria and École Polytechnique, France
Laurent D'Orazio University of Rennes, IRISA, France
Stefania Dumbrava ENSIIE, France
Daniela Grigori LAMSADE, France
Francesco Guerra University of Modena e Reggio Emilia, Italy
Mirian Halfeld Ferrari University of Orléans, LIFO, France
Benoit Lange Inria, France
Zoltan Miklos University of Rennes, IRISA, France
Philippe Rigaux CNAM, Paris, France
Claudia Roncancio Grenoble INP, France
Pierre Senellart ENS, France
Olivier Teste IRIT, France
Karine Zeitouni University of Versailles-St-Quentin, France

Contents

Multi-objective Test Recommendation for Adaptive Learning 1
   Nassim Bouarour, Idir Benouaret, and Sihem Amer-Yahia

Handling Dropouts in Federating Learning with Personal Data
Management Systems 37
   Julien Mirval, Luc Bouganim, and Iulian Sandu Popa

ANTM: Aligned Neural Topic Models for Exploring Evolving Topics 76
   Hamed Rahimi, Hubert Naacke, Camelia Constantin, and Bernd Amann

A Data-Driven Model Selection Approach to Spatio-Temporal Prediction 98
   Rocío Zorrilla, Eduardo Ogasawara, Patrick Valduriez, and Fábio Porto

Optimistic Data Generation for JSON Schema 119
   Lyes Attouche, Mohamed-Amine Baazizi, and Dario Colazzo

Author Index 153

Multi-objective Test Recommendation for Adaptive Learning

Nassim Bouarour[1(✉)], Idir Benouaret[2], and Sihem Amer-Yahia[1]

[1] CNRS, Université Grenoble Alpes, Saint-Martin-d'Hères, France
{nassim.bouarour,sihem.amer-yahia}@univ-grenoble-alpes.fr
[2] EPITA, Lyon, France
idir.benouaret@epita.fr

Abstract. Upskilling is a fast-growing segment of the education economy [31]. Yet, there is little algorithmic work that focuses on crafting dedicated strategies to reach high-skill mastery. In this paper, we formalize ADUP, an iterative upskilling problem that combines mastery learning [49] and Zone of Proximal Development [7]. We extend our previous work [9] and design two solutions for ADUP: MOO and MAB. MOO is a multi-objective optimization approach that relies on Hill Climbing to adapt the difficulty of recommended tests to three objectives: learner's predicted performance, aptitude, and skill gap. MAB is a meta approach based on Multi-Armed Bandits to learn the best combination of objectives to optimize at each iteration. We show how these solutions are combined with two common learner simulation models: BKT (KT-IDEM) [47] and Item Response Theory (IRT) [53]. Our simulation experiments demonstrate the necessity of leveraging all three objectives and the need to adapt the optimization objectives to the learner's progression ability as MAB offers a higher mastery rate and a better final skill gain than MOO.

1 Introduction

The rapid growth in new learning opportunities e.g., MOOCs, tutorials, and community-based discussion forums, is shifting attention to online skill improvement. Upskilling that is occurring outside of formal offerings is a fast-growing segment of the educational economy [31,45]. Moreover, nowadays, learners engage in self-directed learning, managing many elements of their own study, which, in turn, often requires working on various learning activities independently with less direct guidance from teachers [22]. Consequently, providing guarantees on the quality of learning outcomes is increasingly difficult in these new bite-sized learning structures as they can lead to the so-called illusion of explanatory depth [55] where learners only acquire a superficial understanding of a topic. Ideally, each learner should receive tests chosen in a such way that the learner's skill progresses. This should account for the learner's ability to resolve tests based on skill and past performance. That is the topic of mastery learning [49] where the focus of instruction is the time required for different learners to acquire the same competencies and achieve the same level of mastery. To the best of our knowledge, our work is the first to propose formalization that encounters mastery learning. This learning strategy is very much in contrast with classic models of teaching where all learners are given approximately the same time to learn. We illustrate that with an example.

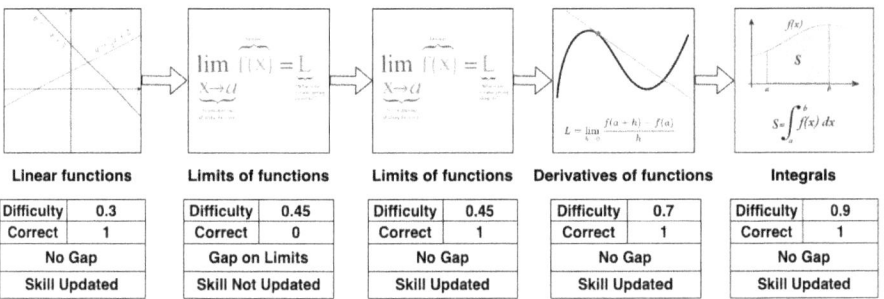

Fig. 1. Example of the process of learning mathematical functions.

Motivating Example. Consider a learner with very basic knowledge in Math who wants to learn mathematical functions. Figure 1 illustrates an example of the learning process. In the beginning, the learner receives tests with a moderate difficulty level of 0.3 for which she provides correct answers. As a result, she incurs no skill gap, and her skill is updated accordingly. This triggers a second step where she is assigned more difficult tests (on limits of functions) on which she fails. In addition to not increasing her skill, she incurs a skill gap. To fill that gap, she is given a second chance with the same type of tests on which she succeeds. Her input is correct and her skill is updated. The same process is repeated, and the learner receives more difficult tests on derivatives and then on integrals. She provides correct results and her skill increases.

Challenges. Our example identifies several challenges. First, we need to determine which k tests to assign to a learner at each iteration. Existing work on recommending tests optimizes the learner's expected performance either by assuming tests with the same difficulty level [49] or by pre-defining the composition of difficulties beforehand (e.g., by alternating test difficulty levels [36]). Indeed, according to learning theories illustrated in Fig. 2, simply relying on the learner's expected performance runs the risk of narrowing down the learner into a zone of "boring" and under challenging tests that do not incur upskilling. To address that, we propose to also account for the learner's aptitude, i.e., the difference between the learner's skill and the test difficulty level. This will encourage selecting tests that challenge the learner (the learnable zone in Fig. 2). Hence, we need to balance expected performance and aptitude. Second, we need to account for the potential skill gap for determining the next k tests. This will motivate the learner to work on her weaknesses and previous failures. To the best of our knowledge, no existing work does so. Third, we need to simulate the learners' performance and devise a skill update strategy after they complete a batch of k tests.

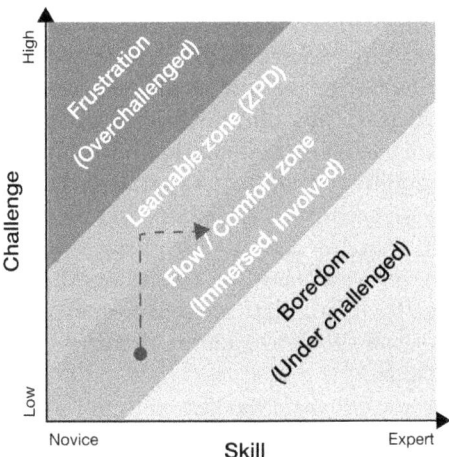

Fig. 2. Illustration of the combination of the Zone of Proximal Development (ZPD) and Flow Theory. In [7], it is shown that learners improve their skills by completing tests that are challenging but not too hard (dotted line).

Contributions. This paper represents an extension of our previous work [9]. We formalize ADUP (Adaptive Upskilling) as an optimization problem where a learner receives k tests that maximize expected performance and aptitude and minimize the accumulated skill gap. The combination of these objectives constitutes the novelty of our formalization.

The main challenge in solving ADUP, is its multi-objective nature. We propose to explore two solutions: a Multi-Objective Optimization, referred to as MUO, and a Multi-Armed Bandits solution, referred to as MAB. MOO is the natural formulation when all dimensions need to be optimized and is addressed by developing a Pareto solution that relies on dominance between k test sets and a *Hill Climbing* [42] heuristic algorithm that finds a subset of the non-dominated solutions [6]. Several variants can be drawn from MOO depending on the different compositions between the objectives. A drawback of MOO is that all variants optimize exactly the same dimensions over all the assigned batches of tests during the whole learning process. It would be desirable to have an approach that learns to find the dimensions to optimize at each iteration. For example, if the learner keeps providing wrong answers to the same tests, favoring the optimization of gap could be more desirable as we need to make sure that the learner successfully completes tests before providing more challenging ones. Therefore, we propose

MAB, a solution that learns which of the three optimization dimensions to optimize at each iteration of k tests. We formalize this approach as a multi-armed bandit (MAB) problem.

Empirical Validation. Our experiments examine the effectiveness of the optimization dimensions on upskilling. To investigate that, we divided our experiments into two parts. In the first one, we examine the impact of our solutions on mastery by simulating the learners' answers as well as the whole learning process. We formulate four research questions: **RQ1.** Is the combination of all optimization dimensions well-adapted for attaining mastery and improving skill gain? **RQ2.** Do different settings of the skill update strategy exhibit different results? **RQ3.** Does the choice of the learner simulation model impact mastery and skill gain? **RQ4.** Does an application of a meta-strategy that chooses to optimize a subset of dimensions at each iteration (MAB), improve mastery achievement? In the second part, we examine the quality of the assigned tests at the next iteration of the learning process based on the interactions of learners with real educational systems. We then formulate a research question **RQ5.** Does the optimization of all the dimensions lead to a more relevant test assignment compared to state-of-the-art baselines?

We use three real-world datasets: MatMat collected from a Czech educational system (matmat.cz)[1], ASSISTment challenge data collected from the ASSISTment platform[2], and, ASSISTment2009 data that contains learners' answers to questions extracted from the same ASSISTment platform between 2009 and 2010. From each dataset, we infer the difficulty levels of the tests based on their features (e.g., type of the test) if they are available. If the features are missing, the inference is based on the correctness rate of the tests of all learners. To simulate learners and predict their probability of providing correct answers, we leverage two models: an extended version of *Bayesian Knowledge Tracing (BKT)* [14] that leverages test difficulties [47] and the *Item Response Theory model (IRT)* [53]. These models capture the learning process of learners and infer the tests that are correctly answered by them. After each iteration, the skill of a learner is updated following an existing approach (i.e., *NCC* [27]) that aggregates the learner's performance based on her N last consecutive answers. For example, when $N = 3$, the learner's skill is updated if she provides three consecutive correct answers on tests with the same difficulty level.

Summary of Findings. For the first part of the experiments, we summarize the findings of each research question. On **RQ1**, we find that combining all objectives performs better than optimizing one or two dimensions only. We show that it yields the highest mastery in fewer iterations. Our results confirm the ZPD and Flow theories [7] and the importance of leveraging aptitude and challenging learners. Moreover, as our skill update strategy is based on *NCC*, we find, on **RQ2**, that MOO is not sensitive to the variation of the value of N. On **RQ3.**, we

---

[1] https://github.com/adaptive-learning/matmat-web/blob/master/data/data_description.md.
[2] https://sites.google.com/view/assistmentsdatamining.

find that the results of **RQ1**, performed on BKT, are generalized when using another simulation model (i.e., IRT). We also observe that MOO offers the highest rate of mastery and optimizing aptitude remains essential. The main difference between the two simulation models is that IRT tends to favor the minimization of gap while BKT favors the maximization of expected performance. On **RQ4.**, we find that choosing automatically the combination of dimensions to optimize at each iteration improves both the skill gain and the percentage of learners that attain mastery. This confirms the need for a meta-strategy that automatically adapts to the learner's skill progression. For the second part of the experiments, we find, on **RQ5.**, that MOO is not only the best in terms of upskilling but it offers the possibility to a larger number of learners to improve their skill compared to recommendation-based (e.g., KNN [56]) and knowledge-based baselines [51, 66](i.e., models that capture the learning and the knowledge of learners). This shows that our optimization dimensions capture with accuracy the evolution of the learning process of a given learner. It also shows that our solution can adapt to different learners.

Organization. In Sect. 2, we define our data model as well as the optimization dimensions we study and we give a formalization of the ADUP problem. Section 3 provides a description of our algorithms. Our extensive experiments are described in Sect. 4. We provide a review of the related work in Sect. 5. We conclude with a summary and discussion on future work in Sect. 6.

2 Model and Problem

We consider a learner $l \in \mathcal{L}$ who follows an iterative learning process for a skill sk. We focus on one skill that has a scalar value. Extending the skill representation to a vector is not straightforward. It requires studying independence between skills or making an independence assumption which may be unrealistic.

At each step, l completes a set of k tests with different difficulty levels for sk. Each test $t \in \mathcal{T}$ has a fixed difficulty d_t. We associate to each learner l a skill value $l.sk$ that either remains the same or increases as the learner successfully completes tests. The initial value of $l.sk$ can be computed from the information the learner fills when joining the system (e.g., by completing an initial set of tests or through a pre-assessment questionnaire). We consider that a learner attains mastery when hey skill value $l.sk$ can not be further improved and is equal to the highest difficulty level.

We aim to formalize a problem where at any given iteration, the learner receives a batch of k tests whose difficulty level is strictly greater than $l.sk$. To define our problem, we formalize dimensions that characterize the learning process of a learner l for a skill sk.

2.1 Expected Performance, Aptitude, and Gap

Expected Performance. It is the expected performance of learner l for a test t. It is based on the similarity of t with successfully completed tests $l.\mathcal{S} \subseteq \mathcal{T}$ by l and is formalized as follows:

$$exPerf(l, t) = sim(t, l.\mathcal{S})$$

Aptitude. It quantifies the difference between a learner's skill value ($l.sk$) and the difficulty level of a test t (d_t). It represents the learner's progression ability for skill sk when assigned tests that are correctly completed. Aptitude is defined as follows:
$$apt(l, t) = d_t - l.sk$$

Gap. It quantifies the distance between the past failed tests of learner l (set $l.\mathcal{F} \subseteq \mathcal{T}$) and the test t and is defined as follows:
$$gap(l, t) = dist(t, l.\mathcal{F})$$

Similarity and distance between tests can be computed in several ways. In our implementation, we use the Euclidean distance between the difficulty levels of tests.

2.2 The ADUP Problem

To achieve skill mastery, we propose an iterative formulation that solves the following problem:

Problem 1 (The ADUP Problem). Given a learner l, with a skill $l.sk$, find a batch $B \subseteq \mathcal{T}$ of k tests to assign to l at iteration i s.t.:

$$\begin{aligned} maximize & \sum_{t \in B} exPerf(l, t) \\ maximize & \sum_{t \in B} apt(l, t) \\ minimize & \sum_{t \in B} gap(l, t) \\ subject\ to\ & |B| = k \end{aligned} \quad (1)$$

3 Solutions

The main challenge in solving ADUP is its multi-objective nature. Scalarization is a common approach that transforms the problem into a single objective whereas optimization dimensions are combined via a weighted linear sum. Another approach is the ϵ-Constraint method where a single objective is optimized and the other objectives are constrained with user-specific values [46]. These methods suffer from the need to fix weights or thresholds, leading to sub-optimal solutions. Therefore, we propose to explore two solutions: a Multi-Objective Optimization, referred to as MOO, and a Multi-Armed Bandits, referred to as MAB.

3.1 Multi-Objective Optimization (MOO)

We propose an approach that finds the Pareto solutions by addressing all objectives at once [6]. To do so, we define a dominance relation between two sets of tests of size k.

We represent the set of all test batches as $C_k = \{B | B \subseteq \mathcal{T}, |B| = k\}$. We define batch dominance $B_1 \succ B_2$ between any two sets in C_k:

Batch Dominance. We say that B_1 dominates B_2 ($B_1 \succ B_2$) iff:

- B_1 is no worse than B_2 for all three objectives.
- B_1 is strictly better than B_2 for at least one objective.

We design a heuristic Algorithm 1, based on [42], to avoid an exhaustive exploration of the search space of possible solutions. It starts by performing *times* iterations where in each it finds an optimal batch of tests (Lines 3 to 7) to avoid local optima. At each iteration, it first generates a random set of k tests. Then it performs *Hill Climbing* to optimize both expected performance and aptitude using Algorithm 2. The Hill Climbing optimizes these two dimensions because they may be related. Indeed, the tests with a high aptitude tend to have lower expected performance (See Fig. 2). The returned candidates are added to the set of results. From this set, only non-dominated candidates are kept (Line 8). Finally, the candidate that yields the lowest gap is chosen (Line 9) and assigned to the learner (Line 10). The skill gap is not directly optimized within Hill Climbing as it is independent of both aptitude and expected performance. The learner's skill is updated after the completion of the test batch (Line 11). Refer to Sect. 4.2 for our skill update strategy. This process is repeated until the learner l masters the skill (i.e., correctly answering the most difficult test).

Algorithm 1: Heuristic MOO that optimizes Aptitude, Expected Performance, and Gap

---

Input: learner l, set of tests $\mathcal{T}$, size k, # repetition *times*

1 **while** *not mastery* **do**
2     $Results \leftarrow \emptyset$
3   **for** n *in* $[1..times]$ **do**
4        $C \leftarrow Random\_candidate(k)$
5        $C^* \leftarrow HCAE(C)$ \\Algorithm 2
6        $Results.Add(C^*)$
7   **end**
8   Keep non-dominated candidates in $Results$
9     $B \leftarrow$ The solution from $Results$ with the lowest skill gap
10   l completes B
11     $l.sk \leftarrow skill\_update(l.sk, B)$
12 **end**

Algorithm 2: HCAE - Hill Climbing for Aptitude and Expected Performance (Called from Algorithm 1)

Input: Batch of k tests B
Output: Optimized batch B^*
1 **while** $True$ **do**
2     $Candidates \leftarrow \emptyset$
3   **for** $test \in B$ **do**
4        $test\_down \leftarrow$ A test with the next lower difficulty
5        $B\_1 \leftarrow B - \{test\} + \{test\_down\}$
6        $test\_up \leftarrow$ A test with the next higher difficulty
7        $B\_2 \leftarrow B - \{test\} + \{test\_up\}$
8        $Candidates.add([B\_1, apt(B\_1), exPerf(B\_1)])$
9        $Candidates.add([B\_2, apt(B\_2), exPerf(B\_2)])$
10   **end**
11   Keep non-dominated candidates in $Candidates$
12   **if** B *dominates all candidates in* $Candidates$ **then**
13   **return** B
14   **end**
15   **else**
16        $B \leftarrow$ A random candidate from $Candidates$
17   **end**
18 **end**

Algorithm 2 is a routine that is called from Algorithm 1 and searches over all the neighbors of the input batch and selects the one that improves aptitude and expected performance. A neighbor of a batch is computed by replacing one and only one test with another test that has either the next higher or next lower difficulty (Lines 3 to 10). If all neighbors are dominated by the current batch, this latter is chosen as the optimized batch. Otherwise, the algorithm replaces the current batch by randomly selecting one from the non-dominated neighbors.

MOO Variants. There are multiple solution variants to ADUP: MOO as described in Algorithm 1; MOEG, MOAG, and MOAE optimize expected performance and gap, aptitude and gap, or aptitude and expected performance respectively; MOG, MOE, and MOA optimize gap only, expected performance only, or aptitude only respectively. Similarly to Algorithm 1, the bi-objective variants are also based on *Hill Climbing* to explore the space of batches and find potential candidates. In the case of MOAE and MOEG, the *Hill Climbing* optimizes expected performance. The condition in Line 9 (Algorithm 1) relates to aptitude for MOAE and gap for MOEG. On the other hand, for MOAG, the *Hill Climbing* optimizes aptitude, and Line 9 remains unchanged.

3.2 Multi-Armed Bandits Algorithm (MAB)

A drawback of the previous solution is that all the variants optimize exactly the same dimensions over all the assigned batches of tests during the whole learning process. However, it would be desirable to have an approach that can learn to find the dimensions to optimize at each iteration. For example, if the learner keeps providing wrong answers to the same tests, optimizing gap solely could be more desirable as we need to make sure that the learner successfully completes these tests before providing more challenging ones. On the contrary, if the learner answers correctly the last batches of tests, it might be better to optimize aptitude so that the learner gets challenged with more difficult tests as she has no gap in her learning process. Therefore, our goal is to design an approach that chooses automatically which of the three dimensions will be optimized at each iteration of k tests. We formalize this approach as a multi-armed bandit (MAB) problem.

The goal of MAB is to verify if a meta approach could be used to address the ADUP problem. The meta approach chooses, at each iteration, an optimization variant of our problem, i.e., bi-objective, or multi-objective optimizations, to generate k tests. We formalize that as a multi-armed bandit problem where each arm corresponds to an optimization variant and the reward r_i, at iteration i, for each variant v is defined as the speed of skill progression:

$$r_{iv} = \frac{\sum_{\forall \text{iterations } j, j < i} \text{skill gain offered by } v \text{ at iteration } j}{\#\text{time the variant } v \text{ was chosen}}$$

At each iteration, the skill progression speed of each arm is computed and the one with the highest cumulated progression is selected. In the case where an arm has never been selected before, its speed is set to zero. The batch of k tests is then generated based on the variant of the chosen arm.

MAB Variants. We implemented different multi-armed bandit strategies [58]: ϵ-GREEDY that chooses randomly an arm (i.e., variant) with an ϵ probability. It chooses the arm with the highest reward with a 1-ϵ probability. THOMPSON Sampling which selects the arm with the highest probability that is learned from previous interactions. The third strategy is the upper confidence bound (UCB) which combines the reward and an uncertainty measure with a confidence degree (c) that balances between exploitation and exploration. Finally, the SOFTMAX strategy relies on Boltzmann distribution that has a parameter (τ) that specifies the randomness of the exploration to choose the optimal arm.

4 Experiments

In this section, we conduct extensive experiments to show the effectiveness of our proposed solutions. We divide our experiments into two parts. In the first part, we compare variants of both MOO and MAB to select the best combination of optimized dimensions to provide the best learning experience to users. We evaluate the overall sequence of assigned test batches from the beginning of the learning process until attaining mastery by simulating the learners' answers. In the second part, we evaluate the quality of the generated batches, at one iteration of the process, based on interactions of real learners. We compare our variants to state-of-the-art adaptive learning and recommendation models. In the following, we first introduce the datasets that we use. We then describe our skill update strategy and finally present the settings and results of each part of the experiments.

4.1 Datasets

We use three real-world datasets that we summarize in the following.

– **MatMat**[3]: The data is collected from a Czech educational system. It is an adaptive practice system for elementary arithmetic tests. The data contains more than 1800 tests from which we infer 42 distinct difficulty levels ranging in $]0,1[$. We assume this order of difficulty level: "divisions" > "multiplications" > "subtractions" > "additions" > "numbers". We consider that all tests for "numbers" have the lowest difficulty (0.13). The difficulty ranges of "additions", "subtractions", "multiplications", and "divisions" are $[0.2, 0.4[$, $[0.4, 0.6[$, $[0.6, 0.8[$, and $[0.8, 1[$ respectively. Within each difficulty range, we assume that multi-digit operations are more difficult than single-digit ones, and tests displayed with visualisations are simpler than directly written tests.
– **ASSISTment** Challenge[4]: The data contains information about the free online tutoring ASSISTment platform[5]. The dataset is composed of school math exercises sampled from the Massachusetts Comprehensive Assessment System (MCAS) containing different types of tests. From the different versions of datasets, we used the ASSISTment challenge one. It contains more than 3000 tests answered by 1709 students. From the data, we selected 10 types of tests (e.g., additions, fractions) and inferred, using the same procedure as MatMat and based on the features of the tests, 26 distinct difficulties.
– **ASSISTment2009**[6]: The data also contains information about the free online tutoring ASSISTment from the 2009–2010 year. We use a subset of the data that was extracted by [35]. The data contains more than 17700 tests answered by more than 4000 students. We classified the tests into different classes and each class has a difficulty level. As the data does not contain tests' features, they are classified by their overall rate of correct answers. We consider the tests with a high rate of correctness the easiest.

4.2 Skill Update and Mastery Achievement

At each iteration and after the completion of a batch B of k tests, we update the skill of learner l as follows:

$$skill\_update(l.sk, B) = max_{sk \in D \cup \{l.sk\}} sk \qquad (2)$$

---

[3] https://github.com/adaptive-learning/matmat-web/blob/master/data/data_description.md.
[4] https://sites.google.com/view/assistmentsdatamining.
[5] https://new.assistments.org.
[6] https://sites.google.com/site/assistmentsdata/home/2009-2010-assistment-data/skill-builder-data-2009-2010?authuser=0.

where D is the set of difficulty values of correctly completed tests for which all tests with lower difficulties were correctly completed.

Table 1. Test completion examples.

Test	Difficulty	Learner Input
t_4	0.35	True
t_5	0.4	False
t_6	0.45	True

To show the intuition of this strategy, we consider a learner with $l.sk = 0.3$ at iteration i. At the next iteration $i+1$, the learner is targeted with $k = 3$ tests t_4, t_5, and t_6 having 0.35, 0.4, and 0.45 as difficulty levels respectively. We consider that the learner correctly answered t_4 and t_6 and failed t_5 (Table 1). Using our strategy, the skill value $l.sk$ is updated to 0.35 (difficulty of t_4). The correct completion of t_6 is not considered as there exists one test (t_5) with a lower difficulty that was wrongly completed. To account for variability in learners' answers, we used the static mastery detection method NCC [27] that updates the skill if the number of consecutive correct answers for a given difficulty level is N. For mastery achievement, we consider that learners attain mastery when their skill can not be further improved and is equal to the highest difficulty level.

4.3 Research Questions

Our goal is to address the following research questions related to the performances of MOO and MAB:

- **RQ1:** Is the combination of all optimization dimensions well-adapted for attaining mastery and improving skill gain?
- **RQ2:** Do different settings of the skill update strategy exhibit different results?
- **RQ3:** Does the choice of the model of learner simulation impact mastery and skill gain?
- **RQ4:** Does an application of a meta-strategy that chooses to optimize a subset of dimensions at each iteration, improve mastery achievement?
- **RQ5:** Does the optimization of all the dimensions lead to a more relevant test assignment compared to state-of-the-art baselines?

We divide these questions into two parts. The first part contains the four first questions where we evaluate the whole learning process of learners (i.e., from the beginning of the process until attaining mastery). To do so, we simulate the learners and their answers. The second part contains only the last question where we evaluate the quality of the assigned test batches on real learners instead of simulating them.

In the following, we first present the models used for simulating the learners. We then present the different metrics, baselines, and experimental settings. We finally report the results to answer each RQ.

4.4 Experimental Settings

Learner Simulation. To answer the first four questions, we need to simulate the answers of the learners. From all the existing simulation models, we rely on these established ones:

- *Bayesian Knowledge Tracing (BKT).* This model simulates learners using an extended version of BKT (KT-IDEM) [47] that takes into account the difficulty level of tests. BKT is a cognitively diagnostic form of assessment that has been recognized as beneficial to learners and instructors [47]. It models the learning process given the chronological sequence and correctness of tests. It infers the knowledge of learners by predicting the probability of learning. In addition, two more probabilities are used to estimate the performance of the learner: Guess and Slip. Guess is the probability of correctly answering a test when the learner does not master the difficulty while Slip is the probability of incorrectly answering a test even if the learner masters the difficulty. If the test is easy, the probability of Guess is high. If the test is hard, the probability of Slip is high as the learners are likely to make mistakes. We use the implementation of [5] in our experiments.
- *Item Response Theory (IRT).* This model simulates learners based on latent factors [11]. The probabilities of the next tests are calculated by applying a sigmoid function and learning a logistic regression to predict responses of learners. One method, AFM [11], infers the probability by characterizing the learner and the difficulty of tests with two distinct parameters. Another method, PFM [48], extends AFM by integrating the number of successes and failures as parameters in addition to previous ones. Other latent models are based on Item Response Theory (IRT) [53], a traditional cognitive diagnosis model [30]. The simplest version [53] predicts a probability of a binary answer (correct/incorrect) by assuming a unique internal parameter for each learner. In addition, it defines tests with one parameter (difficulty) [52], two parameters (the number of correct answers and difficulty) [8], or three parameters (probability of correct answer) [34]. In our experiments, we used this last method based on the implementation of [59]. The reason is that compared to other latent models, it incorporates the probability of guessing in addition to the difficulty and the number of correct answers.

BKT and IRT are structurally different as BKT captures the learning as a chronological process while latent models do not capture the temporal dimensions. They are trained differently as BKT uses the Expectation Maximization (EM) algorithm [5] and IRT uses Adam [59]. Despite these differences, both BKT and latent models infer the probability of correct answers and simulate the learning by capturing similar concepts: the difficulty of tests, the level of learning, and the probability of guessing the correct answers.

Variants. In our experiments we used two types of variants. In the simulation part, we compare MOO and its variants described in Sect. 3.1 as well as MAB and its variants described in Sect. 3.2. We recapitulate them in Table 2. In the real-learners part (i.e., **RQ5**), our goal is to verify whether our dimensions, introduced in Sect. 2.1, capture well the knowledge of the learners. We then use the MOO solution that combines all three of them. We define two variants of the solution: the original MOO, presented in Sect. 3.1, that assigns batches that only contain tests with higher difficulties and MOO_BEG that assigns batches of tests for which the difficulties are lower than the skill level of the learner.

Table 2. Recapitulation of the variants of our solutions

	Variant	Description
MOO	MOO	optimize expected performance, aptitude and gap
	MOAE	optimize expected performance and aptitude
	MOAG	optimize aptitude and gap
	MOEG	optimize expected performance and gap
	MOE	optimize expected performance only
	MOA	optimize aptitude only
	MOG	optimize gap only
MAB	ϵ-GREEDY	choose randomly an arm with an ϵ probability
	SOFTMAX	rely on Boltzmann distribution
	THOMPSON	sample the arm with the highest probability
	UCB	combine the reward and uncertainty with a confidence degree (c)

Baselines. As we want to simulate the whole learning process, we consider ALTERNATE, a state-of-the-art approach that assigns a random set of k tests whose difficulty levels alternate in a round-robin fashion: k easy then k medium then k hard tests [36].

Moreover, to verify the quality of our solution on real learners, we compare it to state-of-the-art methods that capture the knowledge of learners and its evolution. Different models [2] were proposed in the literature. These models differ in their assumptions about the way they represent and quantify the learners' knowledge. They also differ in their strategy of tracking the learning progression. As they can capture the knowledge state of the learners, they were also used in the literature for test assignments and recommendations [26]. In this work, we consider a wide range of methods covering both recent and mostly used ones. We give a brief description for each one:

– KNN [56]: k nearest neighbors. This baseline is commonly used for recommendations. As adaptive upskilling can be seen as a recommendation problem, we

used this method as a baseline to verify how standard recommenders behave for upskilling. KNN finds a predefined number of tests that are similar to the learning path of the learner using cosine similarity. The k tests that have a high similarity are the assigned ones.
- BKT [47]: Bayesian Knowledge Tracing. As explained in more detail in Sect. 4.4, this baseline models the knowledge of learners given the chronological learning path of the learner. It computes the probability of correctly answering a test. For a given learner, the batch of tests that maximizes these probabilities is assigned.
- IRT [53]: Item Response Theory. We also introduced this baseline in Sect. 4.4. It is a popular solution that estimates the performance of learners by learning a logistic function. For a given learner, the batch of tests that maximizes the performance is assigned.
- MCD [59]: Matrix-Factorization Cognitive Based. This method applies the matrix factorization method to education-related data. It embeds both learners and tests and represents the link between them in a latent space. The model learns the latent space based on the learning path of the learner.
- NCDM [63]: Neural Cognitive Diagnosis Model. It is a recent model that incorporates neural networks to learn the learner-test interactions. It projects learners and tests to factor vectors and captures knowledge relevancy and proficiency. It leverages multi-neural layers to output a predicted score of the correctness of a test by a given learner.
- DKT [51]: Deep Knowledge Tracing. It is an extension of the original BKT. It relies on Recurrent Neural Networks (RNNs) [32] to model the learners' process and predict their probability of correctly answering a test. The tests are ranked based on their probability.
- DKVMN [66]: Dynamic Key-Value Memory Network. It uses a key-value memory network [37] to capture the learners' knowledge state and its evolution. It is composed of two matrices: the key that stores the representation of test difficulties and the value that stores the knowledge level of the learner. In this model, the key memory is static while the value one is dynamic (updated after each iteration). The model produces the probability of correctly answering a test.
- SAKT [44]: Self-Attentive Knowledge Tracing. This model adds an attention mechanism to the original knowledge tracing models. It uses the mechanism proposed by [61] to learn attention matrices. It also incorporates multiple attention heads. Each attention matrix learns the importance of a test in the past interactions of the learned in predicting the correctness of the current test. It also predicts the probability of answering correctly a test.

We can group the selected baselines into three categories: Recommendation-based baselines (KNN), Traditional knowledge tracing (BKT, IRT, MCD), and Advanced knowledge tracing (NCDM, DKT, DKVMN, and SAKT). In the last category, we selected the models based on their internal structure of capturing the knowledge and its evolution (Deep Learning, Recurrent Learning, Key-Value Network, and Attention Mechanism).

Metrics. We report (1) the average skill gain i.e. the difference between the last and first skill values for all simulated learners, and (2) the average skill progression i.e. the average skill evolution from iteration to iteration. To better understand this first experiment, we examine (3) the percentage of learners who attained mastery and (4) the average number of iterations required to attain mastery. Finally, we compute (5) the average time each variant takes to generate a batch of k tests.

We want to report the quality of the generated test batches for real learners. For this reason, we evaluate both the logic [67] of the assigned test batches wrt the previous interactions of the learner and the potential knowledge gained when correctly performing these tests. To capture the logic of the assigned batch, we rely on relevance metrics. These metrics are learner-based and evaluate the proportion of the tests that were assigned and were truly performed by the learner in real world (i.e., relevant tests). We then report (6) Precision i.e. the proportion of assigned tests that are relevant, (7) Recall i.e. the proportion of relevant tests that are assigned, and (8) F-Score i.e. a combination between Precision and Recall that captures the balance between them. Finally, we report (9) the percentage of learners who improved their skills under each variant or baseline.

Parameters. In the first four questions, we evaluate the overall sequence of batches assigned to the learners from the beginning until attaining mastery. We set the maximum number of iterations to attain mastery to 500. We vary the value of k in $\{3, 5, 10, 15, 20\}$ and the number of simulations, i.e., learners, in $\{50, 100\}$. We only report results of 100 simulations. Results on other settings are similar.

On the other hand, in **RQ5**, we evaluate the quality of assigned batches for real learners at one iteration of the process. This iteration is chosen based on the split of the data. In fact, as selected baselines need training, we split our data in a learner-wise fashion, i.e., the split is done for every learner, chronologically according to the provided timestamps so that 70% of the data constitutes the training set and the remaining 30% is assigned to the test set. In the case where the timestamps are not provided, the split is random.

We rely on this GitHub repository[7] for the implementation of NCDM, DKT, DKVMN, and 3AKT. We used the same training configurations and the same values of the hyperparameters of the models. We also rely on [59] for the implementation of IRT and MCD). We refer the reader to our GitHub repository[8] for our complete results and code for reproducibility.

4.5 RQ1: Impact of Optimizing All Dimensions

To verify the impact of optimizing all dimensions, we use BKT (KT-IDEM) [47] and assume $N = 1$ in the skill update strategy. We consider two datasets: MatMat

---

[7] https://github.com/hcnoh/knowledge-tracing-collection-pytorch/tree/main.
[8] https://github.com/AdaptiveUpskilling/AdUp.git.

and ASSISTment2009 challenge. We consider two settings: fixed initial skill value and variable initial skill value.

- **Fixed initial skill value.** We assume the same fixed initial skill value for all learners and consider that learners attain mastery when their skill equals the highest difficulty level. We set the initial value to the lowest difficulty level in our simulated data.

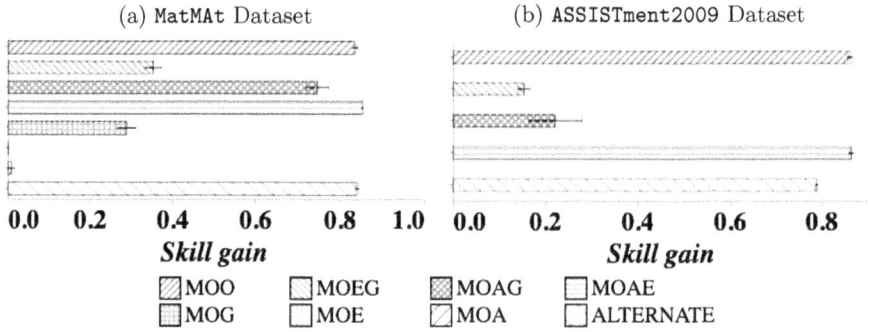

Fig. 3. Average skill gain for each variant with a fixed initial skill.

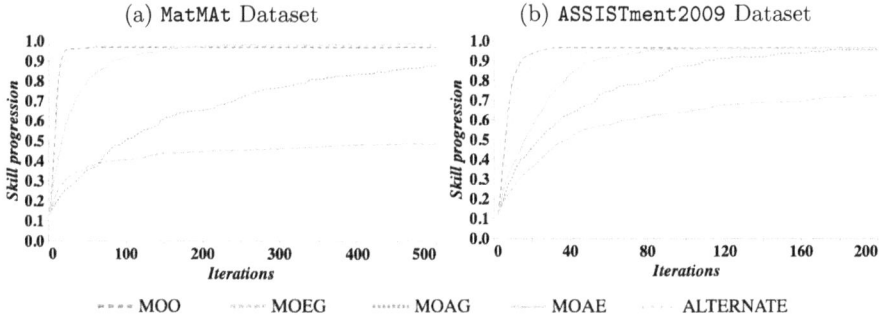

Fig. 4. Skill progression as a function of # iterations with a fixed initial skill.

Skill Gain and Progression. Figure 3 reports the average skill gain. We observe that MOO and MOAE produce the highest average skill gain for both MatMat and ASSISTment2009 datasets. Surprisingly, ALTERNATE seems to also produce a high skill gain outperfoming in both cases MOAG and MOEG. To elucidate that, we plot Fig. 4 to examine the average step-wise skill progression. Here again, we observe that MOO and MOAE result in the fastest upskilling with a clear advantage for the former. MOAG is slower but still faster than MOEG. This reinforces our initial assumption that optimizing for all three objectives at once yields the best results. It also shows that alternating task difficulties does yield good skill gain and progression. Therefore, in the next experiment, we examine whether ALTERNATE compares favorably to MOO and MOAE in terms of achieving skill mastery.

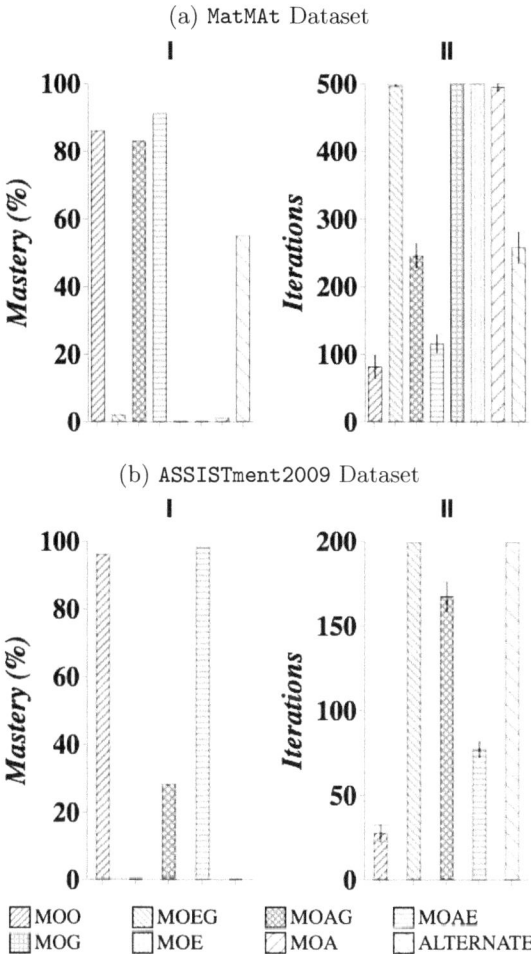

Fig. 5. **(I)** Percentage of learners who attain mastery - **(II)** Average number of iterations to attain mastery.

Mastery. Figure 5(I) reports the number of times each variant attained mastery. One can see that while ALTERNATE reaches a reasonable mastery level in MatMat ($\approx$59%), it is much lower than MOO, MOAG and MOAE ($\approx$90%). On the other hand, one can see that on ASSISTment2009, MOO and MOAE remain effective while ALTERNATE mastery rate is null. This clearly confirms that aptitude plays a central role in attaining mastery as all variants that optimize it offer higher mastery rates than ALTERNATE. Hence, while alternating test difficulty levels do achieve good skill gain (Fig. 3) and skill progression (Fig. 4) performances, it is capped in terms of mastery level since it does not explicitly optimize aptitude. We can also observe that single-objective variants rarely attain mastery. This experiment confirms our initial assumptions: MOE assigns tests that are similar to the ones the learner completed correctly, thereby staying within the under-challenging zone [62]. MOA assigns tests that are too difficult and that keep the learner in a frustration zone [62].

Figure 5(II) shows the average number of iterations to attain mastery for each variant. One can observe that ALTERNATE attains mastery in a similar number of iterations as MOAG in MatMat but has a lower rate of mastery. Nevertheless, it is quicker than all single-objective variants. As explained before, these variants narrow the learners into zones where their skill value does not evolve while ALTERNATE offers more challenging batches that allow learners to attain mastery more often. However, simulated learners under ALTERNATE are able to correctly complete difficult tests but are unable to do so for the most difficult tests. Finally, the figure shows that MOAE attains a slightly higher mastery level than MOO in both datasets, but it is clearly outperformed by MOO in terms of the number of iterations needed to achieve that mastery level.

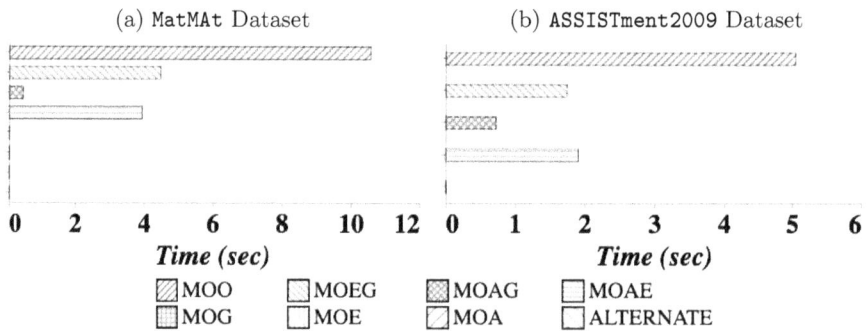

Fig. 6. Average time for generating one batch.

Response Time. Time experiments have shown that single-objective variants are obviously the fastest to generate a batch of k tests (Fig. 6). MOO has the worst time average as it has to optimize three objectives ($\approx$10 s for MatMat and $\approx$5 s for ASSISTment2009). MOAE would be a good candidate since it runs faster than MOO. However, MOO does better than MOAE on skill progression and on the average number of iterations needed to attain mastery. Therefore, we will need to focus on improving response time for MOO in future work.

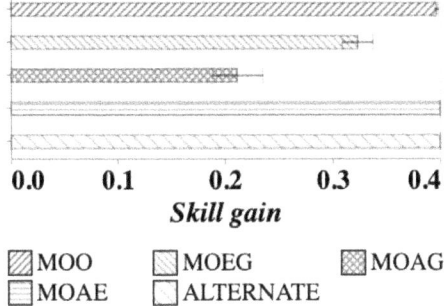

Fig. 7. Average skill gain with variable initial skills on `MatMat`.

- **Variable initial skill value.** We study the case where learners have different initial skill values and consider that a skill is mastered when the skill gain attains a fixed value. We set the value to 0.4 as it is the highest skill gain in common between all learners. We only present results on `MatMat` as similar observations are made on `ASSISTment2009`. We report only bi-objective and `MOO` variants in addition to `ALTERNATE` as we showed already that single-objective solutions are inefficient.

Skill Gain and Progression. From Fig. 7, which reports the average skill gain for all variants, we note that similarly to the case of a fixed initial skill value, `MOO`, `MOAE`, and `ALTERNATE` offer the highest skill gain that is equal to the maximum value (0.4). Figure 8 also generalizes previous results by showing that `MOO` and `ALTERNATE` skill progressions are the fastest followed by `MOAE`.

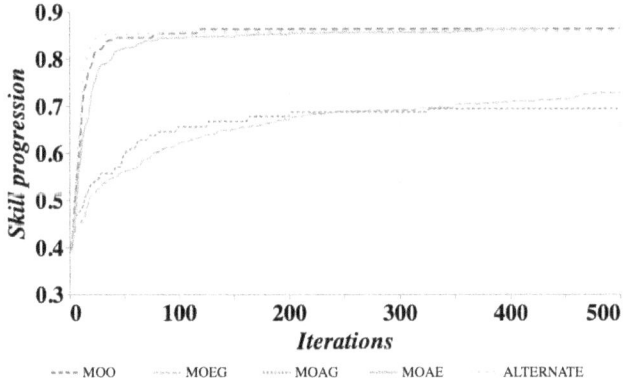

Fig. 8. Skill progression as a function of # iterations with variable initial skills on `MatMat`.

Mastery. Figure 9 shows the percentage of mastery attained by each variant as well as the number of iterations needed to attain mastery. One can confirm that, despite a smaller gain value to attain mastery, optimizing aptitude is still necessary as MOEG is the worst performer for the number of iterations and the second worst for mastery. We also see that ALTERNATE has comparable results to MOO and MOAE which confirms that it is capped in terms of mastery. Obviously, we can see that all variants attain mastery more often and in fewer iterations than when initial skills are fixed. This is due to the fact that in the latter, learners must achieve a much higher skill gain to attain mastery.

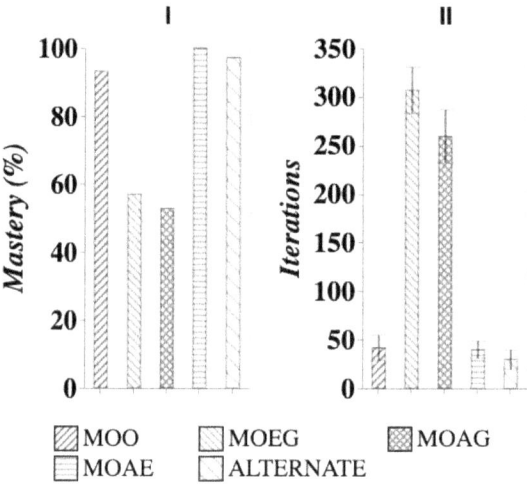

Fig. 9. (**I**) Percentage of learners who attain mastery - (**II**) Average number of iterations to attain mastery using variable initial skills on MatMat.

Findings. This experiment shows that combining all objectives yields the highest skill gain which permits a higher mastery in fewer iterations independently of the initial skill value of the learners. It also shows that challenging learners and optimizing aptitude is beneficial to attain mastery. These results were observed on two different datasets MatMat and ASSISTment2009 that have different characteristics. These results also confirm the ZPD and Flow theories [7] and show the importance of leveraging aptitude and challenging learners.

4.6 RQ2: Impact of Changing the Settings of the Skill Update

We report skill and mastery results by further challenging the learners during the skill update. We increase the value of N, the number of consecutive correct answers, to $N = 3$. We report results in the case where the initial skill is fixed for both MatMat and ASSISTment2009.

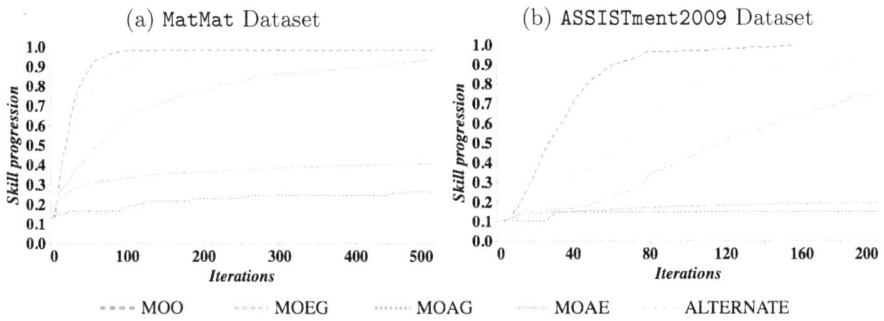

Fig. 10. Skill progression as a function of # iterations with $N = 3$.

Skill Gain and Progression. The average skill gain is similar to the one reported in Fig. 3 where ALTERNATE is comparable to MOO and MOAE. MOO is the best variant while MOAE is the second best. Figure 10 shows the average progression of the skill on both datasets. We see that the progression is slower than the one presented in Fig. 4 because the skill gain is slow. This is intuitive as learners have to answer correctly $N = 3$ tests of the same difficulty level to see their skill updated while previously one correct answer was enough. The second observation is that MOO is still the best variant with a clear advantage compared to ALTERNATE and MOAE. This means that MOO is less affected by the different values of N than other variants. To be sure of this conclusion, we study the Mastery rate and number of iterations under the $N = 3$ constraint.

Mastery. The results, in Fig. 11, show that more than 90% of learners attain mastery under MOO while less than 70% achieve it under ALTERNATE and MOAE on MatMat. On the other hand, 99% of learners attain mastery under MOO on ASSISTment2009. We also see a small decrease in the mastery rates of MOAG, MOEG and ALTERNATE. Results also show that MOO is the fastest as it offers learners fewer iterations to reach the highest difficulty level on both datasets. These results confirm that MOO is not affected by different settings of the skill update strategy and is the best variant.

Findings. This experiment finds that MOO is not sensitive to varying different settings of the skill update strategy and that holds for all datasets.

4.7 RQ3: Impact of Changing the Learner Simulation Model

We report the results of the same metrics using a different learner simulation. We used item response theory (IRT) as explained in Sect. 4.4. We report results, on MatMat only, where the initial skill value is similar for all learners. Similar results were observed when the initial skill value was different from one learner to the other.

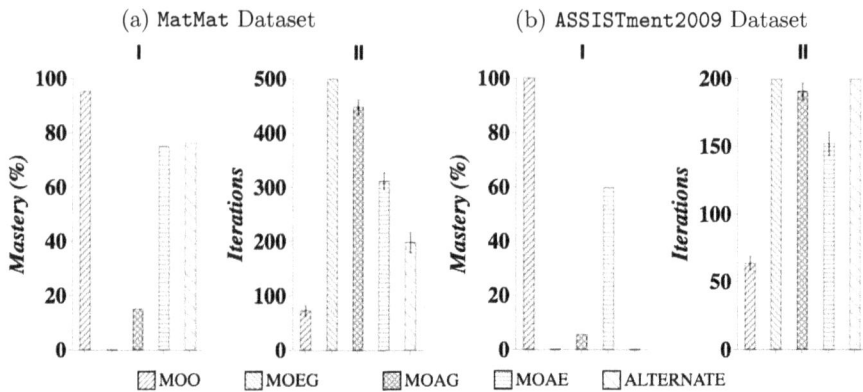

Fig. 11. (I) Percentage of learners who attain mastery - **(II)** Average number of iterations to attain mastery with $N = 3$.

Skill Gain and Progression. Figure 12 reports the average skill gain for the variants that performed well previously. We observe that MOO and MOAG along with ALTERNATE produce the highest skill gain. One can also note that, similarly to the case of KT-IDEM, MOEG is the worst bi-objective variant. The main reason is that the test batches of MOEG do not challenge the learners as MOEG does not optimize for aptitude.

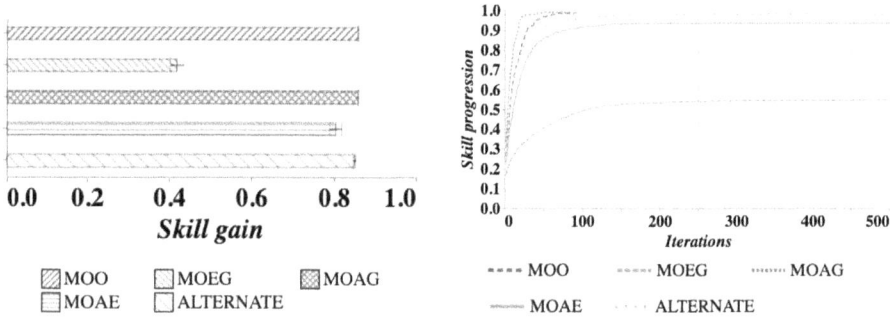

Fig. 12. Average skill gain using IRT on MatMat.

Fig. 13. Skill progression using IRT on MatMat.

Figure 13 shows the step-wise skill progression. We observe that MOO and MOAG are the fastest in terms of upskilling outperforming ALTERNATE which was equivalent to MOO under BKT. One can also observe that MOEG is the slowest. Next, we compare these variants in terms of mastery.

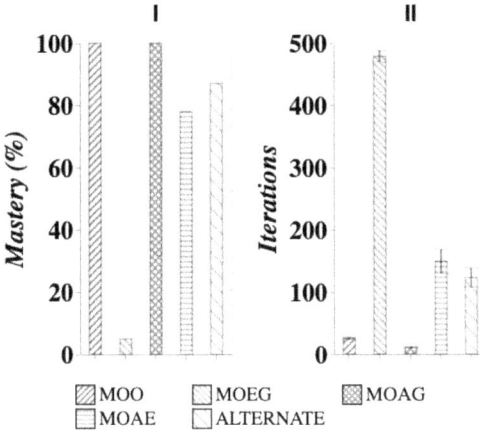

Fig. 14. (**I**) Percentage of learners who attain mastery - (**II**) Average number of iterations to attain mastery using IRT on `MatMat`.

Mastery. Figure 14 shows the average rate of mastery achieved by each variant as well as the average number of iterations to attain it. From the figure, we observe that the state-of-the-art alternating solution (`ALTERNATE`) achieves a high mastery level ($\approx$80%) but it is clearly outperformed by `MOO` and `MOAG`. This experiment confirms that aptitude is required to attain a high rate of mastery as we see that `MOEG` is the worst variant. It attains mastery in $\approx$7% of time. From this figure, we can also observe that `MOAE` is outperformed by `MOAG` while it was better under the BKT model. A hypothetical explanation is related to the internal design of both methods. BKT formalizes the learning process as a hidden Markov model where test completion is viewed as a chronological sequence and where the different parameters are learned using the correctness of tests. In this case and intuitively, learner performances on recently assigned tests appear to be more influential than older tests while in the case of IRT, and because of the absence of time dimension, all performances have the same weight. Usually, as the gap is related to earlier tests, IRT seems to give more attention to it than BKT. Another possible explanation is that BKT tends to overestimate the importance of failure as reported in [48]. In that work, it was observed that BKT tends to predict worse performance after an incorrect answer. Based on that, one can make a hypothesis that BKT is negatively biased towards gap in contrast to latent factors models.

Results from Fig. 14(II) are inversely proportional to the ones depicted in Fig. 14(I). Variants with the highest mastery percentage are the quickest to attain it. Inversely, the variants that attain a lower rate of mastery are the slowest. This indicates with more evidence that `MOO` is the best variant.

Findings. This experiment finds that IRT generalizes the results of KT-IDEM. In this case, we also observe that `MOO` offers the highest rate of mastery. Optimizing aptitude remains essential as `MOEG` is the worst variant. Despite the differences between BKT and IRT, one can explain their similar results with the fact that they both assume a guessing probability of correct answers and characterize tests by their difficulties. They both infer the correctness probability by approximating the knowledge of the learner based on previous correct answers (See Sect. 4.4). In addition, prior work [23,54] has shown that these models exhibit similar prediction accuracy. However, from these results, we see that the main difference between the two learner simulation models is that IRT tends to favor gap as MOAG is comparable to MOO while BKT favors expected performance as MOAE was the second best. We believe that further research needs to perform a more detailed comparison to understand why BKT and IRT offer the same predictions.

4.8 RQ4: Impact of the Meta-strategy

We seek to verify whether choosing automatically a subset of learning dimensions to optimize at each iteration improves mastery and skill progression compared to optimizing fixed dimensions throughout the process. We implemented the four MAB strategies described in Sect. 3.2 and tested them with $N = 1$ as we showed in **RQ2.** that the value of N has no impact on the performances of our solution. We also assume a fixed initial skill value as we showed in **RQ1.** that variable initial skill exhibits similar performances. We used MatMat and ASSISTment2009 for this experiment.

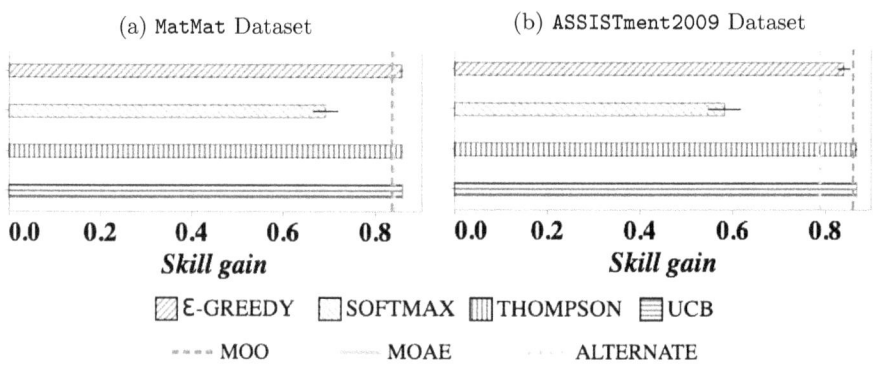

Fig. 15. Average skill gain using MAB strategies.

Skill Gain and Progression. Figure 15 shows the skill gain offered by the different MAB strategies. The lines represent the skill gain attained by learners under MOO and ALTERNATE. One can see that UCB and THOMPSON strategies slightly improve skill gain compared to MOO and ALTERNATE. We also see that SOFTMAX is the worst strategy showing that probability-based MAB is not adapted to this context in both datasets. Similar results can be observed in terms of skill progression in Fig. 16. We can see that UCB is as fast as MOO. We can also see that

THOMPSON has a better progression than ALTERNATE and MOAE, especially after a few iterations. Finally, one may note that ϵ-GREEDY has a similar progression than MOAE on the MatMat and a slightly slower on the ASSISTment2009. A possible explanation of these results is that ϵ-GREEDY takes a longer time to converge than the other MAB variants as it explores more especially at the beginning of the process. Moreover, we explain the outperformance of UCB by the fact that this variant converges quickly and always finds the right dimensions to leverage at each iteration.

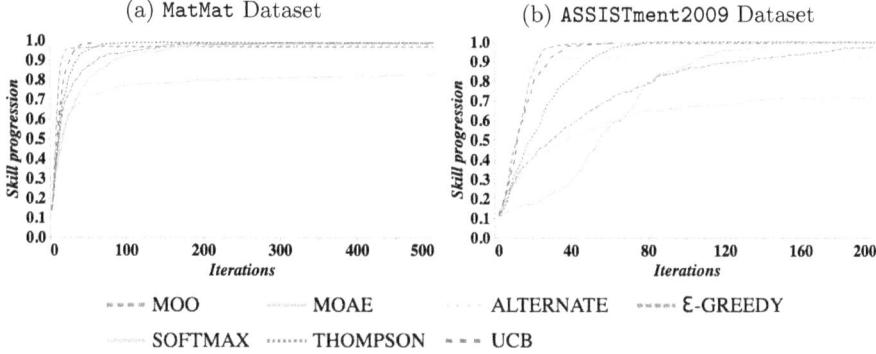

Fig. 16. Skill progression of learners using MAB strategies.

Mastery. Figure 17 shows the percentage of learners that attained mastery and the average number of iterations to achieve that. The lines represent the results of MOO, MOAE, and ALTERNATE. One can see that UCB is the best performer and outperforms all other MAB strategies as well as previous variants for mastery on both MatMat and ASSISTment2009. It also achieves that in fewer iterations. We can also see that THOMPSON attains more mastery than ALTERNATE, MOAE, and MOO in both datasets. It is also better than the two first ones but is equivalent to MOO in terms number of iterations in ASSISTment2009. In addition, ϵ-GREEDY variant is slightly outperformed by MOO in ASSISTment2009 data but is better in terms of mastery and iterations on the second one (MatMat). Finally, even if SOFTMAX is outperformed in terms of skill gain and progression, it achieves higher mastery and in a lower number of iterations than ALTERNATE in both datasets. These findings add more evidence to the skill gain and progression results and confirm our previous assumption that selecting the dimensions to optimize during the learning process is better than optimizing fixed dimensions. However, this involves choosing the right multi-armed bandit variant. In our result, UCB was the best one.

Response Time. Time experiments have shown that MAB strategies are faster than MOO to generate the test batches but are slower than bi-objective variants. The reason is that, during the whole learning process, MAB strategies optimizes

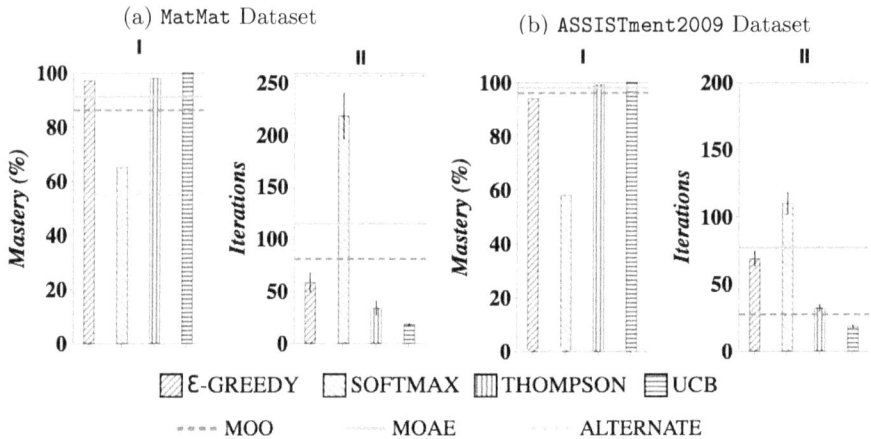

Fig. 17. (**I**) Percentage of mastery attained - (**II**) Average number of iterations to attain mastery using MAB strategies.

fewer objectives than MOO. In some iterations, MAB leverages only two objectives. So they are faster in generating the test batches. On the other hand, MAB strategies optimize more objectives than bi-objective variants. In some iterations, MAB optimizes for all objectives. So they are slower in generating the batches.

Insights on Combining Dimensions. One can see that bi-objective and multi-objective variants are a special case of a MAB strategy where only one arm is available and chosen. Based on that, one can ask the question of whether the policies of MAB are relying on only one or two variants, for example, they leverage both best variants MOO and MOAE. To answer that, we examine the policies output by MAB on MatMat.

First, we examine the overall proportions of the selection of each variant in each strategy. The results show that the best strategies (the ones exhibiting the highest mastery rates in lower iterations) UCB and THOMPSON, exhibit a more uniform use of each variant. For example, in UCB each multi-objective variant is selected $\approx 25\%$ of the time. In contrast, we see that SOFTMAX, the worst strategy, relies mainly on two variants, MOEG with $\approx 84\%$ and MOAE with $\approx 13\%$ of the time. This may be the reason for its underperformance. Another interesting insight is that ϵ-GREEDY selects MOO just 9% of the time. This also explains why ϵ-GREEDY has a higher number of iterations and a slightly slower skill progression.

Analyzing these proportions in more detail showed that UCB is more stable and less noisy in selecting the different variants across all simulations. For example, by calculating the standard deviation of MOO selection proportions we found that UCB has the lowest value (≈ 0.04) while SOFTMAX has the highest one (≈ 0.4). This means that the choice of MOO in UCB is similar from one simulation to another while for SOFTMAX this choice looks more random and noisier.

We now examine the veracity of the hypotheses we made in Sect. 3.2. We assumed that after failing tests, it is more desirable to optimize gap. We also assumed that after obtaining successful answers, aptitude is optimized. Our results show that all strategies tend to leverage gap, in the next two iterations, after learners fail to increase their skill value. For example, UCB and THOMPSON optimize gap 77% and 72% of the time after wrong answers while SOFTMAX does the same 67% of the time. Similarly, our simulations show that UCB, THOMPSON, and ϵ-GREEDY optimize aptitude after successful tests more than 75% of the time, while it is no more than 58% for SOFTMAX. These results also provide insights on why SOFTMAX under-performs compared to the other strategies.

Findings. This experiment finds that choosing automatically the dimensions to optimize at each iteration improves the rate of mastery and the number of iterations needed to achieve it. This justifies the use of a meta-strategy to learn the best combination of objectives to optimize at each iteration.

4.9 RQ5. The Quality of the Generated Test Batches

In this section, we want to verify whether our behavioral dimensions: Expected Performance, Aptitude, and Gap, capture well the needs of real learners. We aim to test if these dimensions define well the knowledge of the learners. We then compare our solution, MOO, to the state-of-the-art adaptive learning solutions [2] presented in Sect. 4.4. We used two of the datasets presented in Sect. 4.1: ASSISTment and ASSISTment2009. In the following, we first present the settings of the experiments and then report the comparison results.

Relevance of Assigned Batches. Tables 3 and 4 report the overall average of Precision, Recall, and F-score over all learners on ASSISTment and ASSISTment2009 respectively. The main observation is that our variants MOO and MOO_BEG are the best solutions for relevance. For example, the Precision of MOO_BEG and MOO is 10% and 9.2% on ASSISTment respectively while the best baseline Precision is 8.8% (i.e., IRT). This means that our solution tends to sufficiently capture the knowledge of the learners and trace its evolution better than all the baselines. More precisely, one can see that MOO_BEG outperforms MOO on both datasets. The reason is that MOO is more restrictive as it only assigns tests that have a higher difficulty than the current skill level of the learner. This shows that, in the real world, learners tend to come back and complete previously assigned tests even if they were mastered. For example, on ASSISTment, 41% of the tests assigned by MOO_BEG were already mastered by learners while this rate is null for MOO.

The second observation is that deep knowledge models tend to have better performances than traditional ones (e.g., BKT, IRT, and MCD). More precisely, we observe that SAKT is the best baseline on ASSISTment2009 and the second best on ASSISTment. It also outperforms all other deep knowledge models. The reason is that it relies on the powerful attention mechanism which mimics cognitive attention and learns the importance and relevance of tests from the input sequence (i.e., the learning process). Our results confirm that the attention mechanism outperforms recurrent models, in an upskilling environment, as it was originally introduced [61] to overcome the drawback of recurrent models (e.g., LSTM [25]). Indeed, recurrent models lose information when embedding the interactions of a long input sequence. Our results also show that models used originally for recommendations can be applied for upskilling. Indeed, we see that KNN has reasonable relevance as it outperforms models like BKT or NCDM on both datasets.

Finally, one can see a drop in performance from ASSISTment to ASSISTment2009. The reason is directly related to the size of the data. Indeed, ASSISTment2009 has more than 17700 tests while ASSISTment contains around 3000 tests. Most importantly, we see that our solutions MOO and MOO_BEG scale well when the number of tests increases.

Knowledge of Assigned Batches. Figure 18 shows the percentage of learners that improved their skill in the next iteration as a function of the average skill (i.e., knowledge) that was gained by these learners on both ASSISTment and ASSISTment2009 datasets. These results are generated by using BKT as a simulator for each learner. From Fig. 18(a), one can see that our solutions MOO and MOO_BEG are outperformed by knowledge tracing baselines in terms of the percentage of learners that improved their skill. One potential explanation for these results is that the knowledge models tend to find a better connection between the embedding of the tests and the performances of the learners. However, despite that, our solutions are the best in terms of the knowledge that these learners gained. We explain this outperformance for the skill gain by the fact that our solution always challenges the learners by maximizing Aptitude while all other baselines tend to optimize only for the Expected Performance. Finally, we see that KNN is the worst performer for both metrics which indicates that recommendation-based solutions tend to assign tests that match the previous sequence the learners interacted with instead of the ones that improve their knowledge.

Table 3. Overall relevance results on ASSISTment dataset

Models		Precision (%)	Recall (%)	F-Score (%)
Recommendation Based	KNN	5.2	1.128	1.854
Traditional Knowledge Tracing	BKT	1	0.232	0.376
	IRT	8.8	1.734	2.897
	MCD	2.8	0.573	0.951
Deep Knowledge Tracing	NCDM	5	1.051	1.737
	DKT	5.2	1.123	1.847
	DKVMN	4.8	0.908	1.527
	SAKT	6.8	1.333	2.229
Ours	MOO	9.2	2.074	3.385
	MOO_BEG	10	**2.151**	**3.54**

Table 4. Overall relevance results on ASSISTment2009 dataset

Models		Precision (%)	Recall (%)	F-Score (%)
Recommendation Based	KNN	0.4	0.01	0.02
Traditional Knowledge Tracing	BKT	0	0	0
	IRT	0.6	0.014	0.027
	MCD	1	0.021	0.042
Deep Knowledge Tracing	NCDM	0	0	0
	DKT	0.6	0.014	0.028
	DKVMN	1	0.026	0.051
	SAKT	1.4	0.037	0.072
Ours	MOO	1	0.024	0.048
	MOO_BEG	**2.2**	**0.055**	**0.107**

On larger datasets (i.e., ASSISTment2009), we see that NCDM is the worst baseline (Fig. 18(b)) while it was the best in the previous data. This baseline does not scale when the number of tests increases. However, we see that our solutions MOO and MOO_BEG are the best for both metrics with an advantage for the former one (e.g., ≈20% of learners improved their skills with an average gain of 0.23). This shows that our methods scale well to the size of the data. Finally, we see that all knowledge tracing baselines have similar performances as was the case in the previous dataset.

Findings. This experiment we performed in this part finds that our solutions are not only the best in terms of skill gain but it offers the possibility to a larger number of learners to improve their knowledge. In addition to that, they assign more relevant test batches according to the previous sequence of learners' interactions. Finally, in this experiment, we found that our solution scales for larger datasets. In conclusion, the dimensions, we defined in this work, tend to capture well the knowledge of the learners and are able to trace their evolution.

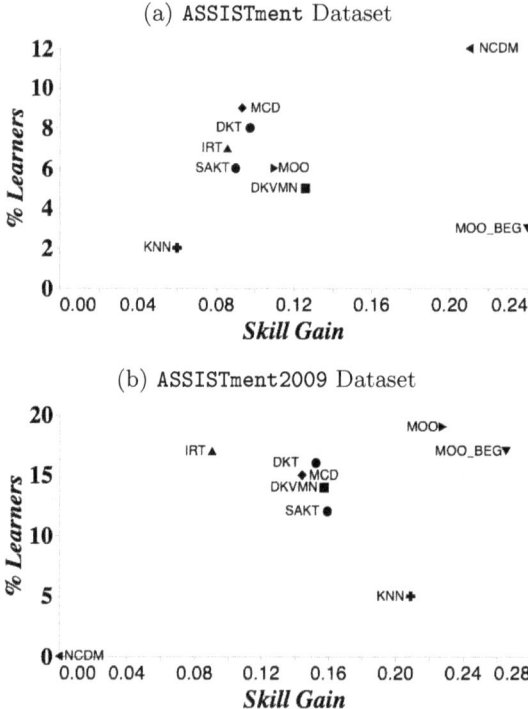

Fig. 18. Average skill gain and average rate of learners who improved their skills for all baselines.

5 Related Work

Education Science. Flow [15] and ZPD [62] theories conceptualize the idea of *experiential learning* [28] that emphasizes the importance of choosing appropriate tests for learners. Flow theory was shown to be effective in the physical world in on-the-job training [18,38]. More recently, it was used in crowdsourcing to compose tasks with different difficulty levels and test the impact on skill improvement and worker satisfaction [36]. *The difference with our work is that the composition of test difficulties is decided beforehand (for instance, by alternating easy and difficult tasks).*

Learner Modeling and Mastery Detection. Many works [49] develop criteria that analyze the sequential interactions of learners and determine if a learner has mastered a skill. The easiest methods are based on simple statistics. For example, *NCC (N Consecutive Correct)* [27] declares mastery if the number of consecutive correct answers exceeds a certain threshold. Another method is *Moving Average* [50] that declares mastery of a type of test if the average of correct answers within a moving window exceeds a threshold. These methods are too simple to capture learners' knowledge of all the skills. For this reason, more sophisticated models were also proposed [2,3]. The two first and most popular models are *Bayesian Knowledge Tracing (BKT)* [14] and *Latent Factor* models [11,48,50].

BKT [14] is a hidden Markov model with 4 parameters: probability that the skill is initially mastered, probability of learning in one iteration, probability of an incorrect answer when the skill is learned (slip), and probability of a correct answer when the skill is unlearned (guess). Many extensions were proposed [2,16,43]. For example, KT-IDEM [47] accounts for the difficulty of each test. Other improvements are made by applying new architecture like Deep Learning techniques. The first extension is Deep Knowledge Tracing (DKT) [51] which uses Recurrent Neural Networks (RNNs) [32], and more precisely LSTM [25], to capture the past performance of the learners chronologically. Other models used key-value memory networks [1,37] to capture these interactions. More models used the concept of BKT with specific data types like text [57] or graphs [41].

On the other hand, *Latent Factor* models are based on logistic regression [11]. They define the performances of the learners using latent parameters and learn them to infer the probability of mastery using a sigmoid function. The most known model is Item Response Theory (IRT) [53] which ignores the chronological aspect of the learner interaction and considers that tests are independent. Other models like Additive Factor Model (AFM) [12] or Performance Factor Analysis (PFA) [48] were proposed by considering other assumptions like the pace with which the learners master a certain skill (or difficulty a difficulty level).

In our work, we leverage some of these models to simulate learners and their answers (e.g., KT-IDEM [47] and IRT [53]). We also used some of these models as baselines to compare the quality of the assigned batches of tests (e.g., DKT [51], and memory networks [37]). Finally, the simplest statistical methods (e.g., NCC) was used in our skill update strategy.

Adaptive Learning. Adaptive learning systems aim to provide an efficient, effective, and customized learning experience for learners by capturing their competencies and interactions with various learning activities and dynamically adapting learning content to suit their individual abilities or preferences [60]. A consistent and growing body of knowledge provides evidence about the effectiveness of adaptive systems compared to classroom teaching or to educational systems that provide instructions and learning activities that are not adaptive [65]. While there are examples of using adaptive learning systems across different disciplines, by and large, they have been most effectively utilized in the context of high school Maths using tools such as ASSISTments [24]. Usually, learner modeling models and knowledge tracing were used for adaptive learning and test assignment [2]. Recent work also combines different types of data and deep learning architectures for adaptive learning [26,33]. In [26], the authors propose to learn a policy based on Reinforcement Learning [4] that optimizes different objectives: Engagement of the learner, Smoothness of the tests, and the trade-off between exploration and exploitation of new tests. The proposed policy outperformed standard baselines like IRT [53], or DKT [51]. Another work by [33] integrates both DKT as a knowledge tracing and a graph-based model as a cognitive navigation for learning path assignment. Their procedure outperforms recommendation-based and simple knowledge models. Finally, [40] also proposes a bi-objective solution for test assignment. The solution maximizes the precision of the assigned tests while minimizing the number of given tests. Their simulated experiments showed the effectiveness of the proposed solution in maximizing the accuracy while minimizing the size of the given tests.

Similarly to some of these works, we aim to solve a multi-objective problem for adaptive learning. On the contrary, and to the best of our knowledge, none of these solutions combines the optimization of expected performance, aptitude, and skill gap to adapt tests to individual learners.

In online labor marketplaces, a few studies focused on the role of task difficulty and workers' ability to complete micro-tasks in improving skills [21], and how affinity between workers can be used to form teams that collaborate to produce high-quality contributions while also improving skills [19].

Usually, such approaches require additional human costs to build training material or give feedback to workers. Additionally, these solutions do not customize test difficulty in recommended tasks.

6 Conclusion and Future Work

We addressed adaptive upskilling following a mastery learning approach. The originality of our approach lies in adapting the difficulty of tests to the learner's predicted performance, aptitude, and skill gap. We proposed two approaches: MOO that directly solves our problem and a MAB that chooses among different optimization variants at each iteration. We tested the impact of optimizing these dimensions on skill progression and mastery achievement. We also tested the impact of different learner simulation models on mastery achievement. Our experiments confirmed that MAB offers a higher mastery rate and a better final skill gain than MOO. They also confirmed that our solution MOO assigns tests with higher quality and accuracy.

For future work, we would like to deploy our solutions so that real learners can interact with them. We may use environments at our university that scaffolds students' activity as they learn to solve exercises or write experimental reports. Experimenting online with real learners will help confirm the findings we exhibit in this work. It will also permit to capture new variables as completion time, reflexion time, or non cognitive metrics (e.g., engagement, motivation) [20].

In addition to that, we aim to extend our formalization by considering additional theories. In fact, there are many learning theories in the physical world, such as situated learning [29] and collaborative learning [10]. One representative of the former is apprenticeship where knowledge is propagated from experts to novice learners based on the principle of *Legitimate Peripheral Participation* [29]. Collaborative learning is also effective in online learning environments like MOOCs, and studies showed that rich interactions such as peer feedback and discussion promote learning [13,17,64].

We also aim to personalize the upskilling experience of learners. We would like to model profiles for learners based on their past performances on the tests of different skills. We may use these profiles to assign tests by either using a clustering method to define their overall ability [39] or applying a collaborative filtering method [56] to focus on the tests that were correctly completed by similar learners.

References

1. Abdelrahman, G., Wang, Q.: Knowledge tracing with sequential key-value memory networks. In: Proceedings of the 42nd International ACM SIGIR Conference on Research and Development in Information Retrieval, pp. 175–184 (2019)
2. Abdelrahman, G., Wang, Q., Nunes, B.: Knowledge tracing: a survey. ACM Comput. Surv. **55**(11), 1–37 (2023)
3. Abdi, S., Khosravi, H., Sadiq, S., Darvishi, A.: Open learner models for multi-activity educational systems. In: Roll, I., McNamara, D., Sosnovsky, S., Luckin, R., Dimitrova, V. (eds.) AIED 2021, Part II. LNCS (LNAI), vol. 12749, pp. 11–17. Springer, Cham (2021). https://doi.org/10.1007/978-3-030-78270-2_2
4. Arulkumaran, K., Deisenroth, M.P., Brundage, M., Bharath, A.A.: Deep reinforcement learning: a brief survey. IEEE Sig. Process. Mag. **34**(6), 26–38 (2017)
5. Badrinath, A., Wang, F., Pardos, Z.: pyBKT: an accessible python library of Bayesian knowledge tracing models. In: Proceedings of the 14th International Conference on Educational Data Mining, pp. 468–474 (2021)
6. Bartolini, I., Ciaccia, P., Patella, M.: Efficient sort-based skyline evaluation. ACM Trans. Database Syst. (TODS) **33**(4), 1–49 (2008)
7. Basawapatna, A.R., Repenning, A., Koh, K.H., Nickerson, H.: The zones of proximal flow: guiding students through a space of computational thinking skills and challenges. In: Proceedings of the Ninth Annual International ACM Conference on International Computing Education Research, pp. 67–74 (2013)
8. Birnbaum, A.: Some latent trait models and their use in inferring an examinee's ability. In: Statistical Theories of Mental Test Scores (1968)
9. Bouarour, N., Benouaret, I., D'Ham, C., Amer-Yahia, S.: Adaptive test recommendation for mastery learning. In: Proceedings of the 2nd International Workshop on Data Systems Education: Bridging Education Practice with Education Research, pp. 18–23 (2023)
10. Bruffee, K.A.: Collaborative Learning: Higher Education, Interdependence, and the Authority of Knowledge. ERIC (1999)
11. Cen, H., Koedinger, K., Junker, B.: Learning factors analysis – a general method for cognitive model evaluation and improvement. In: Ikeda, M., Ashley, K.D., Chan, T.-W. (eds.) ITS 2006. LNCS, vol. 4053, pp. 164–175. Springer, Heidelberg (2006). https://doi.org/10.1007/11774303_17
12. Cen, H., Koedinger, K., Junker, B.: Comparing two IRT models for conjunctive skills. In: Woolf, B.P., Aïmeur, E., Nkambou, R., Lajoie, S. (eds.) ITS 2008. LNCS, vol. 5091, pp. 796–798. Springer, Heidelberg (2008). https://doi.org/10.1007/978-3-540-69132-7_111
13. Coetzee, D., Lim, S., Fox, A., Hartmann, B., Hearst, M.A.: Structuring interactions for large-scale synchronous peer learning. In: Proceedings of the 18th ACM Conference on Computer Supported Cooperative Work & Social Computing, pp. 1139–1152 (2015)

14. Corbett, A.T., Anderson, J.R.: Knowledge tracing: modeling the acquisition of procedural knowledge. User Model. User Adap. Inter. **4**(4), 253–278 (1994)
15. Csikszentmihalyi, M.: Beyond Boredom and Anxiety: The Experience of Play in Work and Games. Jossey-Bass (1975)
16. David, Y.B., Segal, A., Gal, Y.: Sequencing educational content in classrooms using Bayesian knowledge tracing. In: Proceedings of the Sixth International Conference on Learning Analytics & Knowledge, pp. 354–363 (2016)
17. Davis, D., Chen, G., Hauff, C., Houben, G.J.: Activating learning at scale: a review of innovations in online learning strategies. Comput. Educ. **125**, 327–344 (2018)
18. De Vin, L.J., Jacobsson, L., Odhe, J., Wickberg, A.: Lean production training for the manufacturing industry: experiences from Karlstad Lean Factory. Procedia Manuf. **11**, 1019–1026 (2017)
19. Esfandiari, M., Wei, D., Amer-Yahia, S., Basu Roy, S.: Optimizing peer learning in online groups with affinities. In: Proceedings of the 25th ACM SIGKDD International Conference on Knowledge Discovery & Data Mining, pp. 1216–1226 (2019)
20. Fischer, C., et al.: Mining big data in education: affordances and challenges. Rev. Res. Educ. **44**(1), 130–160 (2020)
21. Gadiraju, U., Dietze, S.: Improving learning through achievement priming in crowdsourced information finding microtasks. In: Proceedings of the Seventh International Learning Analytics & Knowledge Conference, pp. 105–114. ACM (2017)
22. Gašević, D., Kovanović, V., Joksimović, S., Siemens, G.: Where is research on massive open online courses headed? A data analysis of the MOOC research initiative. Int. Rev. Res. Open Distrib. Learn. **15**(5), 134–176 (2014)
23. Gong, Y., Beck, J.E., Heffernan, N.T.: Comparing knowledge tracing and performance factor analysis by using multiple model fitting procedures. In: Aleven, V., Kay, J., Mostow, J. (eds.) ITS 2010. LNCS, vol. 6094, pp. 35–44. Springer, Heidelberg (2010). https://doi.org/10.1007/978-3-642-13388-6_8
24. Heffernan, N.T., Heffernan, C.L.: The assistments ecosystem: building a platform that brings scientists and teachers together for minimally invasive research on human learning and teaching. Int. J. Artif. Intell. Educ. **24**(4), 470–497 (2014)
25. Hochreiter, S., Schmidhuber, J.: Long short-term memory. Neural Comput. **9**(8), 1735–1780 (1997)
26. Huang, Z., et al.: Exploring multi-objective exercise recommendations in online education systems. In: Proceedings of the 28th ACM International Conference on Information and Knowledge Management, pp. 1261–1270 (2019)
27. Kelly, K., Wang, Y., Thompson, T., Heffernan, N.: Defining mastery: knowledge tracing versus n-consecutive correct responses. In: Student Modeling From Different Aspects, p. 39 (2016)
28. Kolb, D.A.: Experiential Learning: Experience as the Source of Learning and Development. Englewood Cliffs (1984)
29. Lave, J., Wenger, E.: Situated Learning: Legitimate Peripheral Participation. Cambridge University Press (1991)
30. Lee, Y.W., Sawaki, Y.: Cognitive diagnosis approaches to language assessment: an overview. Lang. Assess. Q. **6**(3), 172–189 (2009)
31. Li, L.: Reskilling and upskilling the future-ready workforce for industry 4.0 and beyond. Inf. Syst. Front., 1–16 (2022). https://doi.org/10.1007/s10796-022-10308-y
32. Lipton, Z.C., Berkowitz, J., Elkan, C.: A critical review of recurrent neural networks for sequence learning. arXiv preprint arXiv:1506.00019 (2015)

33. Liu, Q., et al.: Exploiting cognitive structure for adaptive learning. In: Proceedings of the 25th ACM SIGKDD International Conference on Knowledge Discovery & Data Mining, pp. 627–635 (2019)
34. Lord, F.M.: Applications of Item Response Theory to Practical Testing Problems. Routledge (1980)
35. Mao, Y., et al.: Learning behavior-aware cognitive diagnosis for online education systems. In: Zeng, J., Qin, P., Jing, W., Song, X., Lu, Z. (eds.) ICPCSEE 2021. CCIS, vol. 1452, pp. 385–398. Springer, Singapore (2021). https://doi.org/10.1007/978-981-16-5943-0_31
36. Matsubara, M., Borromeo, R.M., Amer-Yahia, S., Morishima, A.: Task assignment strategies for crowd worker ability improvement. Proc. ACM Hum. Comput. Interact. **5**(CSCW2), 1–20 (2021)
37. Miller, A., Fisch, A., Dodge, J., Karimi, A.H., Bordes, A., Weston, J.: Key-value memory networks for directly reading documents. arXiv preprint arXiv:1606.03126 (2016)
38. Mincer, J.: On-the-job training: costs, returns, and some implications. J. Polit. Econ. **70**(5, Part 2), 50–79 (1962)
39. Minn, S.: BKT-LSTM: efficient student modeling for knowledge tracing and student performance prediction. arXiv preprint arXiv:2012.12218 (2020)
40. Mujtaba, D.F., Mahapatra, N.R.: Multi-objective optimization of item selection in computerized adaptive testing. In: Proceedings of the Genetic and Evolutionary Computation Conference, pp. 1018–1026 (2021)
41. Nakagawa, H., Iwasawa, Y., Matsuo, Y.: Graph-based knowledge tracing: modeling student proficiency using graph neural network. In: IEEE/WIC/ACM International Conference on Web Intelligence, pp. 156–163 (2019)
42. Omidvar-Tehrani, B., Amer-Yahia, S., Dutot, P.-F., Trystram, D.: Multi-objective group discovery on the social web. In: Frasconi, P., Landwehr, N., Manco, G., Vreeken, J. (eds.) ECML PKDD 2016. LNCS (LNAI), vol. 9851, pp. 296–312. Springer, Cham (2016). https://doi.org/10.1007/978-3-319-46128-1_19
43. Ostrow, K., Donnelly, C., Adjei, S., Heffernan, N.: Improving student modeling through partial credit and problem difficulty. In: Proceedings of the Second 2015 ACM Conference on Learning@ Scale, pp. 11–20 (2015)
44. Pandey, S., Karypis, G.: A self-attentive model for knowledge tracing. arXiv preprint arXiv:1907.06837 (2019)
45. Panel, H.E.S.: Final report–improving retention, completion and success in higher education (2017)
46. Papadimitriou, C.H., Yannakakis, M.: On the approximability of trade-offs and optimal access of web sources. In: Proceedings 41st Annual Symposium on Foundations of Computer Science, pp. 86–92. IEEE (2000)
47. Pardos, Z.A., Heffernan, N.T.: KT-IDEM: introducing item difficulty to the knowledge tracing model. In: Konstan, J.A., Conejo, R., Marzo, J.L., Oliver, N. (eds.) UMAP 2011. LNCS, vol. 6787, pp. 243–254. Springer, Heidelberg (2011). https://doi.org/10.1007/978-3-642-22362-4_21
48. Pavlik, P.I., Jr., Cen, H., Koedinger, K.R.: Performance factors analysis–a new alternative to knowledge tracing. In: AIED 2009 (2009)
49. Pelánek, R., Řihák, J.: Experimental analysis of mastery learning criteria. In: Proceedings of the 25th Conference on User Modeling, Adaptation and Personalization, pp. 156–163 (2017)
50. Pelánek, R.: Application of time decay functions and Elo system in student modeling. In: Proceedings of the Educational Data Mining, January 2014, pp. 21–27 (2014)

51. Piech, C., et al.: Deep knowledge tracing. In: Advances in Neural Information Processing Systems, vol. 28 (2015)
52. Rasch, G.: Probabilistic Models for Some Intelligence and Attainment Tests. ERIC (1993)
53. Reckase, M.D.: Unidimensional item response theory models. In: Multidimensional Item Response Theory. SSBS. Springer, New York (2009). https://doi.org/10.1007/978-0-387-89976-3_2
54. Rollinson, J., Brunskill, E.: From predictive models to instructional policies. International Educational Data Mining Society (2015)
55. Rozenblit, L., Keil, F.: The misunderstood limits of folk science: an illusion of explanatory depth. Cogn. Sci. **26**(5), 521–562 (2002)
56. Sarwar, B., Karypis, G., Konstan, J., Riedl, J.: Item-based collaborative filtering recommendation algorithms. In: Proceedings of the 10th International Conference on World Wide Web, pp. 285–295 (2001)
57. Su, Y., et al.: Exercise-enhanced sequential modeling for student performance prediction. In: Proceedings of the AAAI Conference on Artificial Intelligence, vol. 32 (2018)
58. Sutton, R.S., Barto, A.G.: Reinforcement Learning: An Introduction. MIT Press (2018)
59. bigdata ustc: EduCDM (2021). https://github.com/bigdata-ustc/EduCDM
60. VanLehn, K.: The relative effectiveness of human tutoring, intelligent tutoring systems, and other tutoring systems. Educ. Psychol. **46**(4), 197–221 (2011)
61. Vaswani, A., et al.: Attention is all you need. In: Advances in Neural Information Processing Systems, vol. 30 (2017)
62. Vygotsky, L.: Zone of proximal development. In: Mind in Society: The Development of Higher Psychological Processes, vol. 5291, p. 157 (1987)
63. Wang, F., et al.: Neural cognitive diagnosis for intelligent education systems. In: Proceedings of the AAAI Conference on Artificial Intelligence, vol. 34, pp. 6153–6161 (2020)
64. Wu, A.S., Farrell, R., Singley, M.K.: Scaffolding group learning in a collaborative networked environment (2002)
65. Xu, Z., Wijekumar, K., Ramirez, G., Hu, X., Irey, R.: The effectiveness of intelligent tutoring systems on K-12 students' reading comprehension: a meta-analysis. Br. J. Edu. Technol. **50**(6), 3119–3137 (2019)
66. Zhang, J., Shi, X., King, I., Yeung, D.Y.: Dynamic key-value memory networks for knowledge tracing. In: Proceedings of the 26th International Conference on World Wide Web, pp. 765–774 (2017)
67. Zhu, H., et al.: A multi-constraint learning path recommendation algorithm based on knowledge map. Knowl. Based Syst. **143**, 102–114 (2018)

Handling Dropouts in Federating Learning with Personal Data Management Systems

Julien Mirval[1,2,3](✉)[iD], Luc Bouganim[2,3][iD], and Iulian Sandu Popa[2,3][iD]

[1] Cozy Cloud, "Le Surena" face au 5 Quai Marcel Dassault, 92150 Suresnes, France
julien.mirval@cozycloud.cc
[2] Inria de Saclay, Campus de l'école Polytechnique, 1 rue Honoré d'Estienne d'Orves, 91120 Palaiseau, France
luc.bouganim@inria.fr
[3] Université de Versailles Saint-Quentin, 45 avenue des Etats-Unis, 78000 Versailles, France
iulian.sandu-popa@uvsq.fr

Abstract. The development and adoption of personal data management systems (PDMS) has been fueled by legal and technical means such as smart disclosure, data portability and data altruism. By using a PDMS, individuals can effortlessly gather and share data, generated directly by their devices or as a result of their interactions with companies or institutions. In this context, federated learning appears to be a very promising technology, but it requires secure, reliable, and scalable aggregation protocols to preserve user privacy and account for potential PDMS dropouts. Despite recent significant progress in secure aggregation for federated learning, we still lack a solution suitable for the fully decentralized PDMS context. This paper proposes a family of fully decentralized protocols that are scalable and reliable with respect to dropouts. We focus in particular on the reliability property which is key in a peer-to-peer system wherein aggregators are system nodes and are subject to dropouts in the same way as contributor nodes. We show that in a decentralized setting, reliability raises a tension between the potential completeness of the result and the aggregation cost. We then propose a set of strategies that deal with dropouts and offer different trade-offs between completeness and cost. We extensively evaluate the proposed protocols and show that they cover the design space allowing to favor completeness or cost in all settings.

Keywords: Secure aggregation · Peer-to-peer · Reliability · Federated learning

1 Introduction

New privacy-protection regulations (e.g., GDPR) and smart disclosure initiatives in the last decade have boosted the development and adoption of Personal Data Management Systems (PDMSs) [3]. A PDMS (e.g., Cozy Cloud [19], Nextcloud, Solid) is a data platform that allows users to easily collect, store, and manage into a single place data directly generated by the user's devices (e.g., quantified-self data, smart home data, photos) and data resulting from the user's interactions (e.g., social interaction data, health, bank, telecom). Users can then leverage the power of their PDMS to benefit from their personal data for their own good and for the benefit of the community [18].

As a result, the PDMS paradigm leads to a shift in the personal data ecosystem since data becomes massively distributed, on the user side. It also holds the promise of unlocking innovative usages. An individual can now cross her data from different data silos, e.g., health records and physical activity data. In addition, individuals can leverage their PDMSs by forming large communities of users sharing their data. This allows, for example, to compute statistics for epidemiological studies or to train a Machine Learning (ML) model for recommendation systems. In this context, it is natural to rely on a fully decentralized PDMS architecture (as opposed to central servers that raise several important issues such as cost, availability and scalability with the number of users), but this also poses new challenges.

Aggregation primitives are essential to compute basic statistics on user data and are also a fundamental building block for ML algorithms. In particular, Secure Aggregation (SA) is a central component of Federated Learning (FL), introduced in [39], as evidenced by the large body of recent work in this area [38]. FL is defined as machine learning massively distributed on edge devices, where each participant holds its own private data. However, to enable such new usages in the PDMS context, we need new solutions adapted to its specificity. First, PDMS users rely on large peer-to-peer systems for data sharing and computations [3,13], thus requiring fully decentralized and scalable aggregation protocols, discarding data centralization on servers. Also, these protocols need to protect user privacy and adapt to varying selectivity (i.e., the consent of relevant participants). Ideally, the proposed protocol should provide an accurate result that takes advantage of the high-quality data available in PDMSs. Efficiency (i.e., protocol latency and total load of the system) is of prime importance given the potentially limited communication speed or computation power of PDMSs. Finally, given the scale of such decentralized aggregation, protocols must also be robust to node dropouts. To summarize, our goal is to design protocols that fulfill the following properties: **fully decentralized and highly scalable**, with the number of participants; **privacy-preserving**, i.e., protecting the confidentiality of the contributed user data; **accurate**, i.e., no trade-off between accuracy and privacy (e.g., like in the data anonymization or differential privacy approaches); **adaptable**, i.e., adapting to a large spectrum of computation selectivity values (reflecting the subset of contributor nodes) and system configurations (network and cryptographic latency); and **reliable**, i.e., handling node dropouts (e.g., failures, voluntary disconnections or unexpected communication delays).

Ensuring these properties altogether is challenging and to the best of our knowledge, the existing distributed SA protocols fail to achieve this objective. On one hand, approaches such as local differential privacy are based on adding noise to protect privacy. This affects accuracy [9] or reliability to dropouts [48] and requires a very large number of participants to reduce the impact of noise which contradicts an adaptive node selectivity (see Sect. 2). On the other hand, despite leveraging different cryptographic schemes in SA for FL [38] (e.g., encryption-based [8,21] or secret sharing-based [12,17,30]), existing solutions employ a similar hybrid architecture wherein one or several highly available and powerful servers aggregate the data supplied by many user devices. Although some solutions consider the case of node dropouts, this applies to client devices and never to aggregation servers [12,17]. In a Peer-to-Peer (P2P) PDMS system, all computations are performed by internal PDMS nodes (i.e., user devices). Hence, the data aggregators and data contributor nodes have the same constraints, i.e., limited computing power and availability. Such nodes cannot be expected to carry out heavy cryptographic operations [12] and can drop out during the computation. Fortunately, the P2P approach allows involving many nodes to perform a computation thus reducing the load on individual aggregators.

A first effort towards SA adapted to P2P systems was made in [42], where we designed a protocol that fulfill the above properties in an ideal setting, i.e., without considering the reliability issue. In [41], we focus on the reliability property, which is difficult to guarantee in a fully-decentralized setting and deserves a detailed study. In addition, although our protocols apply to SA in general, we chose to study the more general case of FL, given its particular interest in the PDMS paradigm. The study of FL is also more challenging due to the potentially large size of the model, which increases the scalability problem. In our experiments, we consider model sizes from very small to very large, thus covering a wide range of use cases (including classical SA). This paper is an extended version of [41] with a detailed state of the art and a deeper dive into the experimental platform used and the reflections that went into it, leading to more extensive experiments and performance graphs.

Our contributions are as follows. We analyze the impact of dropouts, be it contributor or aggregator nodes, on the other properties of an SA protocol designed for a P2P PDMS system. Node dropouts have a direct impact on accuracy (i.e., a single failure can make the final computation result useless) and on efficiency (i.e., it can introduce large latency). From this analysis, we derive the precise requirements of a reliable protocol and show that in a fully-decentralized context, reliability also introduces a tension between result completeness (i.e., the percentage of initial contribution in the final result, despite dropouts) and computation cost. We introduce the necessary building blocks to deal with these requirements. Then, we propose a variety of execution strategies offering different trade-offs between completeness and cost and allowing to cover a wide spectrum of dropout rates, contributor selectivity or trained model sizes. Our extensive experimental evaluation shows that the proposed strategies cover well the design space allowing to favor completeness or cost in all settings.

The rest of this paper is organized as follows. We discuss the related work w.r.t. the required properties in Sect. 2. We introduce, in Sect. 3, the considered architecture and threat model. Section 4 reminds the main design principles proposed in [42] and then introduces, as a starting point, a straw-man SA protocol which efficiently computes the required aggregation assuming an ideal world (i.e., there are no node dropouts). This allows to highlight the challenges induced by reliability issues. Section 5 presents the necessary building blocks to addresses the reliability related challenges. In Sect. 6, we propose four SA strategies that leverage those building blocks and allow for different trade-off between result completeness and aggregation cost. We extensively evaluate the proposed strategies in Sect. 7 and conclude in Sect. 8.

2 Related Work

Secure aggregation (SA) has been an intense research area since many years leading to several approaches: SA based on (local) differential privacy, anonymity-based SA, SA based on Trusted Execution Environments (TEE) and SA based on cryptography. However, these solutions are not adapted to our decentralized context and fail to cover all the required properties listed above.

2.1 Differential Privacy-Based Secure Aggregation

The concept of differential privacy, initially proposed in [22], centers on the premise that computations on two nearly identical datasets, differing by just one element, should yield results that vary minimally, within a small margin ϵ. Traditional differential privacy models relied on a central authority to collect and process data, ensuring privacy. However, this dependency on a trusted central entity led to the development of Local Differential Privacy (LDP), introduced in [33], which has its roots in the randomized response technique, formulated in [69]. It allows individuals to modify their data locally before sharing. Existing works address problems such as ML [9], FL [1,66,70], marginal statistics [73] or basic statistics based on range queries [16].

Despite the practical applications of differential privacy in various fields, as demonstrated in [20,24], designing algorithms that strike a balance between data utility and privacy remains challenging. The inherent noise addition in LDP however accentuates the tension between utility and privacy protection since it requires more noise for the same level of protection as with classical DP [2]. Consequently, LDP is typically more suited to scenarios involving an extremely large number of participants to counterbalance the noise and where noisy data is acceptable.

Gossip-based protocols, known for their scalability, decentralization, and reliability, typically do not safeguard user privacy. In recent studies [9,53], participants preserve privacy by sharing differentially private models or introducing noise in initial iterations and canceling it over time. Indeed, the absence of a central party of LDP fits well into fully decentralized networks like gossip. However, these techniques have a negative impact on utility and do not take advantage of high-quality PDMS data.

2.2 Anonymity-Based Secure Aggregation

One method to process user data while preserving their privacy involves applying anonymization techniques before computation. This involves uniquely altering the data to prevent the extraction of personal information, yet still enabling computations. Techniques like k-anonymity [62], l-diversity [37], and t-closeness [34] have been proposed to anonymize datasets that contain information from multiple individuals. This allows for the analysis of such datasets without compromising the privacy of the users involved. However, this cannot be used in networks of PDMS because it could introduce privacy risks when grouping data from multiple nodes or if only applying it locally, or it will require strong data degradation.

Mixing techniques have also attracted a lot of interest from the research community. The *Encode-Shuffle-Analyze* framework [11] proposes a way to enable traditional database analysis on anonymous data. [27] proposes a summation protocol working with only a few anonymized messages. However, these techniques do not support nodes dropping out.

A line of work [13,35,36] achieves targeted statistical queries on decentralized networks of PDMS. It focuses on privacy-preserving targeting of participants in a context where small aggregates are computed. Because it only targets a sample of the target population for privacy reasons, this framework is not usable to include data from all relevant participants. Moreover, it neglects the impact of the size of the aggregate because it looks at statistical queries, not machine learning models.

Anonymity-based SA focuses on the anonymization of data pertaining to multiple users, intending to conceal each user into the masses. However, this necessarily implies a degradation of user data in case of failures and is incompatible with our objectives of resilience and maximizing the usage of the high-quality data stored by PDMS.

2.3 Trusted Execution Environment-Based Secure Aggregation

Secure hardware has been used for a long time, but generally for extremely limited scopes. With the recent advances of *Trusted Execution Environments* (TEEs), new hopes have been given of widening the applications of secure hardware. TEEs are hardware solutions allowing to run a program on a machine in an isolated way. This isolation is characterized by a protection against tamper of said program and against observation of the processed data including from privileged software such as the operating system. Recently, TEEs have been included inside many different models of processors embedded in smartphones, computers and servers, opening new possibilities of ensuring security on a potentially insecure device. Intel Software Guard Extensions (Intel SGX) [18] and Arm TrustZone [5] are the leading examples of TEEs. TEEs enhance security in computing by creating an isolated processing environment [54] by encrypting and keeping track of the integrity of the RAM pages used by the code [28]. While being executed inside the TEE, the isolated code can produce a proof of its identity (i.e., attestation) certifying that it is running inside a genuine TEE. Secure hardware on the user side could be used for tasks such as SQL aggregation [65], spatio-temporal aggregation [55], or even to speed-up FL [32] with asynchronous aggregation [45].

However, the security of TEEs is not perfect and several side-channel attacks have been showed. The first type of side-channel attacks target the inner mechanisms of TEEs [15,57], breaking data confidentiality. Some [15,56] even extract the attestation key of the TEE, which enables forging fake attestations. Another type of side-channel attack exploits the design of TEEs, using the untrusted rich execution environment running alongside the TEE to extract information manipulated by programs running in the TEE [14,58]. Research work has been carried out to mitigate the impact of these attacks [25,47]. While this shows that some security flows of TEEs can be corrected, the fallibility of TEEs cannot be ignored.

Ultimately, none of the existing solutions fully meet the requirements for SA using TEE in a PDMS environment. This environment must deal with the inherent challenges of a fully decentralized system, such as inevitable aggregator dropouts and the necessity for accurate FL. Current solutions rely too heavily on the assumption that TEEs are effectively tamper-proof when past research suggests that risks still exist and could lead to catastrophic privacy leaks if exploited.

2.4 Cryptography-Based Secure Aggregation

Cryptographic solutions for SA have been proposed for nearly three decades. The early solutions were designed for wireless sensor networks [49], but the field has recently taken off again to meet the needs of federated learning. A recent survey [38] discusses about forty works for FL, grouped in either encryption-based SA or MPC-based SA.

Encryption-Based Secure Aggregation

Encryption-based SA can be done using various schemes, including additively homomorphic encryption and masking. It involves encrypting private data using cryptographic keys, making the data unusable for individuals who do not have the proper key.

Additively Homomorphic Encryption (AHE) is a property of some cryptography systems which enables computations to be carried out directly on the ciphertext instead of having to decrypt it beforehand, preserving data confidentiality. They were first described in [52] as follows: Given a cryptosystem where E is the encryption function and D the decryption function, this system is said to be additively homomorphic if $E(m_1 + m_2) = E(m_1) \oplus E(m_2)$ where m_1, m_2 are messages. AHE enables highly secure computations but generally comes at a great cost penalizing efficiency. Some schemes have reported up to ten times worse performances [72] compared to the plain text alternative. Moreover, some key management issues can arise which are difficult to manage in the presence of dropouts.

Masking consists of aggregating masks on top of private data, such that the data is kept private unless the aggregation reaches a point where masks cancel each other out. It is one of the key tools used in SecAgg [12,59,60]. Masks are generally much cheaper to generate compared to additively homomorphic cyphertexts and are therefore easier to scale. Nonetheless, handling dropouts implies managing masks and their removal, which can be impractical in a peer-to-peer context.

Multi-Party Computation-Based Secure Aggregation

Unlike encryption-based SA, Multi-Party Computation (MPC)-based SA does not use cryptographic keys to preserve data confidentiality but instead uses private messages sent to other participants according to specific protocols. Best-known members of this category include garbled circuits or computations on secret shares.

Garbled circuits transform computations into logical circuits that are then "garbled" by each party so that they can privately exchange how the circuit should be evaluated without revealing their private inputs. The problem of schemes based on garbled circuits (e.g., [8,21]) is that they require a lot of messages, which prevents scaling to large sets of participants and does not handle failures efficiently.

Computations on secret shares involve transforming private data into smaller pieces of secret that independently reveal nothing but can be recomposed into the original value. Using mathematical properties of these secret shares similar to homomorphic encryption enables the creation of more complex systems (e.g., [26,30]). Some secret sharing schemes are natively redundant, making handling dropouts easier. In some cases, computations on secret shares are analogous to SA using masking.

We can see that Encryption-based SA, with techniques like additively homomorphic encryption and masking, provides robust security by allowing computations on encrypted data. However, these methods often come with trade-offs in terms of efficiency and complexity in key management. On the other hand, MPC-based SA offers an alternative approach, leveraging garbled circuits and computations on secret shares to maintain data confidentiality without the use of cryptographic keys. Despite their potential for scalability and handling dropouts, these methods also face challenges in terms of communication overhead and computational complexity.

Given the scarcity of resources in large networks of PDMS compared to what centralized service providers can offer, encryption-based SA appears as a better alternative. Indeed, the communication overhead introduced by MPC-based SA do not scale well to the number of participants that we envision in our context.

More specifically, we favor masking techniques over homomorphic encryption as it reduces costs even further, by avoiding the use of expensive cryptographic operations.

2.5 Conclusion

In this section, we delved into the various secure aggregation techniques relevant to PDMS environments. Differential privacy-based SA works with noisy private data to protect individual data points but struggles with the trade-off between data utility and privacy. Anonymity-based SA, while useful in some contexts, can lead to data degradation and can be challenging to apply in PDMS networks. Trusted Execution Environment-based SA leverages secure hardware solutions but is not without its security vulnerabilities. Lastly, MPC-based SA offers interesting solutions in terms of security but faces challenges in efficiency, scalability, and handling node dropouts caused by the large communication overhead it introduces.

In summary, encryption-based SA, and particularly masking, stands out as the most fitting technique for PDMS networks, opening the way for solutions that are not only secure and privacy-preserving but also practical in terms of resource utilization and scalability. This explains its popularity in federated learning. However, existing methods using masks rely on high-end highly available aggregation servers which is not feasible in PDMS networks, and perform poorly for a large number of participants and dropouts (i.e., masks, although less costly to manage compared to MPC-based SA, must nonetheless be carefully handled to obtain accurate results).

In a nutshell, the existing secure aggregation methods cannot be applied in a decentralized PDMS setting for two reasons: (i) scalability – a PDMS is not a high-end server that could deal with thousands of connections and related crypto operations, making the existing solutions not scalable with a large number of participants; and (ii) reliability – similar to the client devices (i.e., the data contributors), the aggregator nodes can drop out making the existing solutions inoperative in our context. Adapting secure aggregation techniques to the architectural specificity and the related constraints of the PDMS context is therefore necessary.

3 System Overview and Threat Model

3.1 System Architecture

P2P Network. We envision a fully distributed P2P system relying only on PDMSs, thus requiring an efficient communication overlay. *Distributed Hash Tables* (DHT) are structured overlays which enable a logarithmic scalability with the number of nodes. Our protocol is currently built on top of the Chord DHT [61]. Each node has an *Id* obtained by hashing a static property of the node and stores a *fingertable* (FT) to route Chord messages. FT is a table with a number of entries equal to the size of the *Id* space in bits. If X is a node *Id*, the ith entry of the FT contains the IP address of the node whose *Id* is closest but lower than $X + 2^i$. Routing is done by searching in the FT the closest entry to the target address and transmitting recursively the message until it reaches its target, with a worse case of $\mathcal{O}(log(N))$ message complexity, where N is the number of DHT nodes.

Computation Model. A model computation can be triggered by any node, i.e., *querier*. The querier broadcasts the computation and each node consents or not to contribute, and in the positive case is called *contributor*. The ratio between the number of contributors and total number of nodes defines the *selectivity* $\sigma \in [0, 1]$. Each node (contributor or not) may be a data processor and is then called *aggregator*. Each contributor trains the model locally for several epochs as described in [39] and sends it to the aggregators. Aggregators produce a new model based on the received contributions. The process can potentially repeat for several iterations.

3.2 Threat Model

As in the majority of SA works [38], we consider the classical *honest-but-curious* threat model, i.e., an attacker can access, but cannot alter, the data manipulated by the attacked nodes (called *leaking nodes*). A PDMS can hold the entire digital life of her owner and thus needs to be highly protected against privacy threats as indicated by recent works [4]. However, we consider that some PDMS owners have succeeded in tampering their PDMS since no security measure is unbreakable. Since attackers may collude and thus, de facto, control more than one PDMS, the worst-case attack is represented by the maximum number of colluding nodes controlled by a single "attacker", i.e., C leaking nodes. Additionally, each PDMS is equipped with a trustworthy certificate supplied by an offline PKI. Thus, any node can verify the authenticity of other participants by checking their certificate. This prevents Sybil attacks (i.e., forging nodes to master a large portion of the system). Finally, secure communication channels (e.g., TLS) are required since attackers can observe the communications between the nodes.

Our protocols should fully protect the confidentiality of the contributors' data and all the intermediary results, with high and tunable probability (see also [13]), the final result not being confidential. Also, we consider that being a contributor for a given computation is not a sensitive information.

Out-of-Scope Attacks. We do not consider the case of an attacker manipulating fully corrupted PDMSs. In a P2P system, such an attacker could perform poisoning attacks of the contributed data [17] or forge false aggregation results [29] with the objective to compromise contributors' input confidentiality by bypassing the SA protocol. A few recent works (e.g., Prio [17], VeriFL [29]) deal with these problems but existing solutions are still limited especially in our context because of their limited scalability or lack of genericity.

4 Straw-Man Protocol

This section summarizes the main design principles proposed in [42] to fulfill the privacy, accuracy, adaptability and scalability properties. It then describes, as a starting point, a straw-man protocol in an ideal world, i.e., assuming there are no dropouts. Finally, it highlights the reliability issue by considering node dropouts and formulates precisely the problem at hand.

4.1 Background

Achieving a scalable aggregation process requires multiple aggregators, naturally arranged in a tree structure (see Fig. 1a) wherein the intermediary nodes are aggregators and the leaves are contributors. The querier obtains the result from the tree root.

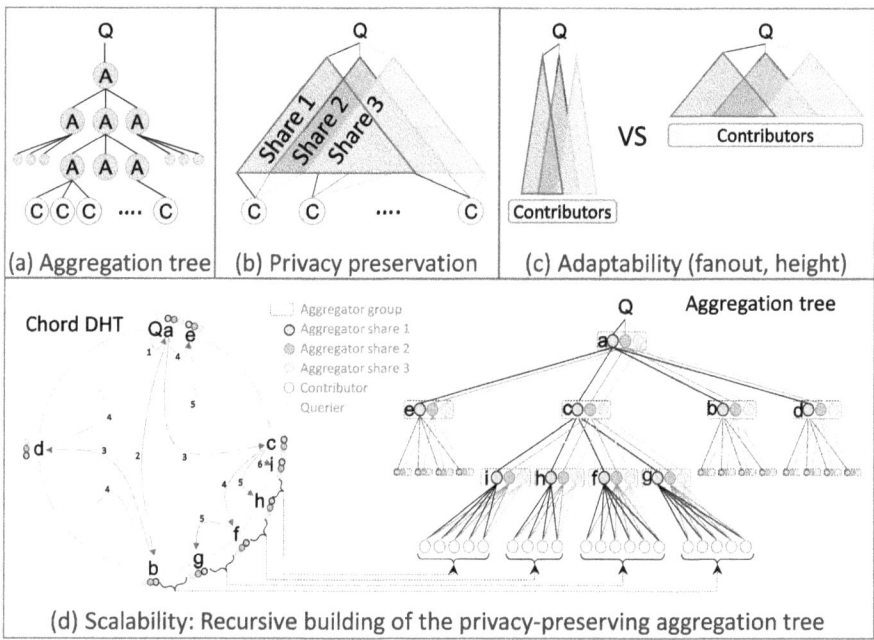

Fig. 1. Building the aggregation tree based on DHT (Color figure online)

Privacy and Accuracy: We use a secret sharing scheme without threshold for data confidentiality. Each contributor splits its private value into s shares, making it impossible to reconstruct the secret unless someone collects all s shares. Considering s shares for each contributor and partial aggregate results leads to build s separate (parallel) aggregation trees with exactly the same structure. This precludes inferences from an attacker on any of the intermediate results (see Fig. 1b). Each ith share has the value $x_i = x/s + \epsilon_i$ such that $\sum_{i=1}^{n} \epsilon_i = 0$, where x is the private value. Thus, shares from different contributors are aggregated separately and if no share is missing (reliability is discussed in Sect. 5), the final result equals the exact sum of all private values and is computed by the querier; hence, our protocol provides, by construction, accurate results.

The number of shares, s, is computed such that the probability to obtain s shares for an attacker, controlling C nodes, is inferior to α, a security threshold (e.g., $\alpha = 10^{-6}$). An attacker could cleverly locate her controlled nodes in the DHT to obtain the s shares of a group. We avoid this attack by reusing the concept of imposed location proposed in [36]: the node Id in the DHT is computed by hashing the public key from the PDMS certificate (see Sect. 3.2). The nodes are then uniformly distributed in the DHT space and the PDMS owner (here the attacker) cannot influence this placement. Consequently, the uniform distribution also applies to leaking nodes and the probability that an attacker controls an entire group is given by $(C/N)^s < \alpha$. Then s is minimal when $s = \lceil \log(\alpha)/\log(\frac{C}{N}) \rceil$.

Obviously, communications between nodes must use secured channels, to protect the integrity and confidentiality of the exchanged data and to ensure the provenance of that data.

Adaptability: The number of aggregators and their arrangement (i.e., the tree fan-out and its height) is tuned as a function of the number of contributors, the communication costs and the processing costs as discussed in [42]. This allows the protocol to always offer near-optimal performance (i.e., aggregation latency) and achieve adaptability w.r.t. the computation selectivity and PDMSs characteristics. Furthermore, our protocol can be configured to offer the desired trade-off between the latency and the total cost of the aggregation, which are conflicting objectives. At one extreme (see Fig. 1c left), a binary tree ($f = 2$) distributes the query load on a maximum number of aggregator groups but increases the communications costs. At the other extreme (see Fig. 1c right), a tree limited to a unique aggregator group ($f = \sigma \times N$) minimizes the communications costs, the total system load but concentrates most of that load on this unique aggregator group that becomes overloaded. Thus, in an "ideal" setting, the height of the tree is chosen to optimize the query latency without impacting too much the total load.

Scalability: The DHT realizes a de facto fully decentralized and efficient architecture for communication between nodes. Building and broadcasting s aggregation trees can be very costly since the trees can be large. We thus employ a divide-and-conquer approach to parallelize the construction and diffusion of the trees and use the finger table structure to minimize communications. Finally, we reduce the knowledge and the diffusion of the trees to the part required to perform the aggregation: a node of an aggregation tree only knows its parent and its children.

To simplify the description of the tree construction, we consider below that each node of the tree is a group of s nodes (see Fig. 1d with $s = 3$). Assuming the querier knows the height h and the fan-out f of the aggregation tree (see above), it starts the tree creation by assigning the whole DHT to its successors. Recursively, each aggregator group in the tree (i.e., parent nodes) is assigned to a DHT region that it will subdivide and delegate to other aggregator groups in that region. When an aggregator oversees a DHT region, it looks for f nodes that are (almost) evenly spaced across the region. The node responsible for finding peers is a parent aggregator, while the selected nodes are child aggregators. Each child then becomes the parent of the region between itself and the next sibling. This process goes on until the height h is reached. At each step, s nodes are selected instead of one. To make this selection efficient, each node in the DHT maintains a cache with the addresses and certificates of the $s-1$ successor nodes that will form the aggregator group. At the last tree level, the tree leaves (i.e., the contributors) are found by using a localized DHT broadcast in the respective region. Figure 1d illustrates this process with three nodes per group (blue, red and green) by using letters to represent a group. The fan-out is 4 and the height is 3 (excluding the querier Q).

4.2 Straw-Man Protocol in an Ideal World

This section details a straw-man aggregation protocol, assuming an ideal world in which there are no dropouts, in order to illustrate the reliability problem. For the sake of simplicity, the presented protocol considers that the aggregation trees were built up to the leaves, but without including the contributors.

Straw-man is detailed in Algorithm 1 by type of nodes considering the computation of the average of a private vector owned by each contributor. Contributors willing to participate establish a secure channel with each aggregator parent and then send shares of their private vector. The aggregators aggregate the received shares and send their results to their parents up to the root. The querier then performs the final aggregation to obtain the result. There are only two types of messages: (i) *Query()* messages containing (1) the query itself (line 25); (2) the sender certificate (line 26), and (3) the receiver parents to whom the shares must be sent (line 28). (ii) Intermediate results under the format ($\overrightarrow{sum}$, $nbContrib$) sent either by contributors (line 28, with $nbContrib = 1$) or (Leaf) Aggregators (lines 12 and 23).

After having broadcasted the query, the leaf aggregators set a *contribution timeout*, computed such that it allows to receive all contributions (line 16). The timeout is computed by considering the time to reach a contributor plus the time to prepare and send a contribution since we consider an ideal world with no delays. While sent in parallel, the contributions are decrypted sequentially by the leaf aggregators, which wait for the processing of any message (line 20) before sending the partial result (line 23). If a node selected as aggregator (leaf or not) in the tree wishes to contribute, it can simply add its private data to the partial aggregate it computes add s to the count of share contributions before sending it to its parent.

4.3 Analysis and Problem Formulation

Although the straw-man protocol is simple and efficient in an ideal world, it can deliver an incorrect result or simply block in the presence of node dropouts.

Algorithm 1: Straw-Man protocol (average computation)

Message definition:
- $IntRes_i(Sum, NbContrib)$: intermediate result of child i.
- $Query()$: ask contributors for their potential contributions.

Input: s: number of shares; f: tree fan-out

Querier Node :

1   on *initialization* do $\overrightarrow{sum} \leftarrow \overrightarrow{0}$; $nbContrib \leftarrow 0$
2   on $IntRes_i()$, $i \in [1..s]$ do
3        $\overrightarrow{sum}$ += $msg.\overrightarrow{sum}$; $nbContrib$ += $msg.nbContrib$
4   if *I received s intermediate results* then
5            $\overrightarrow{result} = \overrightarrow{sum}/(nbContrib/s)$ /* average */

Aggregator Nodes :

6   on *initialization* do $\overrightarrow{sum} \leftarrow \overrightarrow{0}$; $nbContrib \leftarrow 0$
7   on $IntRes_i()$, $i \in [1..f]$ do
8        $\overrightarrow{sum}$ += $msg.\overrightarrow{sum}$; $nbContrib$ += $msg.nbContrib$
9   if *I received f intermediate results* then
10   if *I want to contribute* then
11             $\overrightarrow{sum}$ += $\overrightarrow{myData}$; $nbContrib$ += s
12   Send $IntRes(\overrightarrow{sum}, nbContrib)$ to my parent

Leaf Aggregator Nodes :

13   on *initialization* do
14        $\overrightarrow{sum} \leftarrow \overrightarrow{0}$; $nbContrib \leftarrow 0$
15   Broadcast the query to the assigned part of the DHT
16   Set a *Contribution Timeout* (to receive all contributions)
17   on $IntRes_i()$, $i \in [1..f]$ do
18        $\overrightarrow{sum}$ += $msg.\overrightarrow{sum}$; $nbContrib$ += $msg.nbContrib$
19   after *Contribution Timeout expiration* do
20   if *there is no more pending messages* then
21   if *I want to contribute* then
22             $\overrightarrow{sum}$ += $\overrightarrow{myData}$; $nbContrib$+=s
23   Send $IntRes(\overrightarrow{sum}, nbContrib)$ to my parent

Potential Contributor Nodes :

24   on $Query()$ do
25   if *I want to contribute* then
26   if *msg.sender is a PDMS (check certificate)* then
27   for $i \in [1..s]$ do
28   Send $IntRes(\overrightarrow{share_i}, 1)$ to $msg.parents[i]$

Indeed, one share of a contributor may not be received because the contributor drops out after sending some shares or because the corresponding message was delayed. In both cases, the result is incorrect. Furthermore, if an aggregator fails before sending its intermediate result, the condition in line 9 will never be true, thus blocking the protocol. A single aggregator dropout is indeed sufficient to thwart a graceful protocol termination since all the ancestors of the dropout node will hang on indefinitely waiting for the data to arrive.

The problem addressed in this paper is to design protocols robust to dropouts, i.e., ensure the reliability property with three complementary goals despite failures and delays:

1. *validity*: the protocol must deliver a correct result;
2. *termination*: the protocol must not block;
3. *completeness*: the protocol should maximize completeness defined as the percentage of the initial contributors actually accounted for in the final result.

Termination and validity are mandatory while maximizing completeness is a desirable objective, but may incur a significant overhead. An ideal protocol should minimize this overhead and maximize the completeness of the result, which are unfortunately conflicting goals. Indeed, maximizing completeness requires synchronization between the parallel aggregation trees and the ability to redo the work done by a dropped out aggregator. In addition, this overhead increases the latency of the protocol which can lead to increased dropouts, with, potentially, a snowball effect.

5 Handling Dropouts

This section proposes solutions to handle dropouts during the aggregation protocol. We first introduce the dropout model and detection. Then, we discuss possible approaches to react to dropouts, guarantee validity and termination.

5.1 Dropout Model and Detection

We consider the most difficult failure model wherein any node can dropout at any moment (i.e., we cannot benefit from graceful disconnections). For simplicity, in all cases, we consider that dropout nodes cannot reintegrate the ongoing computation after a dropout. When a dropout node recovers, it reintegrate the DHT or can participate in new queries.

We consider that the dropout probability is the same for any contributor or aggregator node. That is, at each time instant (e.g., every second) during the protocol execution, every node can dropout with some fixed probability, thus, the longer the protocol duration, the higher the risk of a dropout and hence the observed dropouts. In this model, there is no way to detect a dropout with certainty; a dropout can only be assumed after a timeout, i.e., node A may presume node B's dropout because A was expecting a message from B and did not receive it after a given timeout.

Let us note that the aggregation trees form a temporary additional overlay on top of the DHT overlay. Hence, to detect dropouts, we use a common DHT mechanism to maintain its consistency, i.e., health check (HC) messages. Specifically, any aggregator is periodically monitored by its parent using HC messages. HC are sent over the secure channels already required to secure the communications in the aggregation tree. HC are equivalent to ping messages so they imply a low network overhead.

5.2 Replacing (or not) a Dropped Out Node

The natural reaction to the dropout of an aggregator A is to trigger a node replacement as follow: the parent P detecting the dropout of node A randomly selects a free node, say R, from its DHT fingertable (i.e., a node that has not been previously selected as aggregator in the current tree) and supplies R the necessary information (e.g., A's children, the members of A's group, A's status, etc.) allowing the node to take the place of A. This information can be easily kept up-to-date by P when A answers P's health check requests. If the dropout occurs before A has received any data from its children, the replacement is cheap, entailing only the creation of secure communication channels between R and P, A's children and the members of A's group. In the other cases (i.e., A has done part or all of its assigned work), the replacement induces a significant overhead (R must require A's children to re-send their data, potentially redo the aggregation and re-send the aggregated data to P). Thus, all the strategies described in Sect. 6 replace any node dropping out before receiving any data while the replacement policy in the other cases depends on the strategy, with an impact on the overhead/completeness trade-off. Obviously, contributors that drop out cannot be replaced. Thus, a synchronization between the parallel aggregation trees must take place to ensure validity if some contributors or some –not replaced– aggregators dropped out.

5.3 Ensuring Validity

This section introduces two complementary mechanisms for ensuring result validity. The first is based on recording and checking the contributors' footprints in each of the parallel aggregation trees. The second uses inter-tree synchronization between aggregators in the same group allowing for contributors' convergence between the parallel aggregation trees.

Check Contributors Footprint (CCF)

Validity is ensured as long as the s last *IntRes* messages (see Algorithm 1), computed by the s parallel aggregation trees and received by Q, contain the secret share contributions of the exact same list of contributors. To this end, we employ a hashing scheme similar to a Merkle Hash Tree [40], i.e., computing incrementally a hash of the identifier lists of the contributors whose shares are aggregated. We add a new field, CF, to $IntRes$ messages which contains, for leaf aggregators, a hash of all contributors IPs that are included in that intermediate result. Contributors thus send $IntRes(MyShare, 1, hash(MyIP))$. Then aggregators, leaf or not, sort the incoming CFs, hash that sorted list to produce their own CF and send it with their intermediate result. The process repeats to all intermediate levels up to the querier Q. Therefore, Q can ensure the production of a valid result iff all the CFs received together with the s intermediate results are equal. Thus CF can be considered as a *version identifier* of a given $IntRes$. CCF allows for efficient detection of inconsistent shares but not for a convergence of those shares. Hence, the lack of even a single share leads to invalidating the entire aggregation, i.e., $completeness = 0\%$.

Inter-tree Synchronization (Sync)

To correct inconsistencies between the parallel aggregation trees, we need a synchronization mechanism to eliminate from the $IntRes$ message the contributions of any contributor that provided less than s secret shares. Note that this can arrive either because of a contributor dropout but also as a result of an aggregator dropout which is not replaced. This synchronization called *sync* can be applied in a blocking or non-blocking manner as described below.

Blocking Sync. A blocking sync is performed between the aggregators in a same group (e.g., the blue, red and green nodes of any group in Fig. 1d), which synchronize their contributing children list to produce an $IntRes$ containing only the data from children nodes in the intersection of those lists. If the children data was synchronized before aggregation, the resulting $IntRes$ is then consistent (i.e., will have the same CF). Each aggregator in a group waits for all its children data and then broadcasts the list of contributing children to the other aggregators in its group. After receiving $s-1$ lists, each aggregator produces through intersection the final list, aggregates the corresponding shares and sends the result to its parent.

Non-blocking Sync. The idea is to allow aggregators to send the aggregated shares up the tree without synchronization with the other $s-1$ leaf aggregators in the group, but just informing them of the actual *Children List (CL)* used to compute the $IntRes$. A leaf aggregator who receives a CL must react in different ways, depending on its own status: (a) if it has not sent any $IntRes$ message, it must ignore the data from children that are not in the received CL; (b) if it has already sent an $IntRes$ with its own CL (CL_{last}), it computes $CL_{new} = CL_{last} \cap CL$. If $CL_{new} \neq CL_{last}$, the aggregator sends a new $IntRes$ message based on CL_{new} data and informs the other aggregators of its group, sending CL_{new}. Thus, if the querier receives inconsistent $IntRes$ messages, it detects it through the CF inconsistency and just has to wait for new $IntRes$ messages that will eventually become consistent.

For a leaf aggregator group (e.g., the f group in Fig. 1d), the sync eliminates the contributors that provided only a part of the s shares. In the upper tree levels, the sync eliminates entire tree branches and therefore, possibly a significant number of contributors (e.g., if c-blue drops out and is not replaced, the sync at the a group in Fig. 1d leads to prune the whole c sub-tree since there are no means to retrieve the blue shares in this sub-tree). Thus, sync operations may hurt completeness. A blocking sync guarantees that all the group aggregators send consistent *IntRes* up the tree and thus potentially entails a lower cost (i.e., both bandwidth and computation cost) than a non-blocking sync wherein aggregators may send several *IntRes* messages (i.e., eventual consistency). However, in strategies that replace dropped out aggregators, a replaced aggregator may require another sync with the members of its group, thus reducing the interest of a blocking sync. Moreover, blocking syncs may increase query latency since any slowdown in one of the parallel aggregation tree will impact the others. Finally, we should underline that a blocking sync does not require CCF if applied in all groups, whereas this is required for non-blocking sync.

5.4 Ensuring Termination

Ensuring query termination is straightforward insofar as dropouts are detected (see Sect. 5.1). The protocol can gracefully terminate if the querier receives a consistent set of s *IntRes*. In this case, the querier 'broadcasts' a termination message (i.e., which is propagated recursively down the s aggregation trees). Depending on the aggregation strategy (see Sect. 6) a dropout can also trigger termination. For instance, in a straw-man-like protocol a single dropout can invalidate the entire result. Hence, on detecting a dropout, an aggregator informs the querier which sends termination to all nodes. Finally, nodes within sub-trees can receive an early termination message (i.e., before the protocol end) following a sync at the sub-tree root group requiring pruning.

6 Aggregation Strategies

Having laid out the main building blocks that we intend to study, we now move to the topic of assembling and composing them to form coherent strategies. Since our building blocks are never a panacea and always present trade-offs between aggregation cost (i.e., the latency, total work, and bandwidth of the protocol) and result completeness, the goal is to provide Pareto-efficient strategies for different cost minimization and completeness requirements. The strategies introduced below are not exhaustive of what can be built using these building blocks and more building blocks can be created to propose new strategies but, based on our experiments, the proposed strategies are the best combination of our building blocks.

We start by introducing the most extreme strategies called *LowCost* and *HighCpl* , which respectively try to minimize costs and maximize the result completeness. Then, using those strategies as starting points, we introduce two trade-off strategies called *Sync&Prune* and *Hybrid* that try to mitigate the weaknesses of extreme strategies. For each strategy, we describe its overall objective,

6.1 LowCost

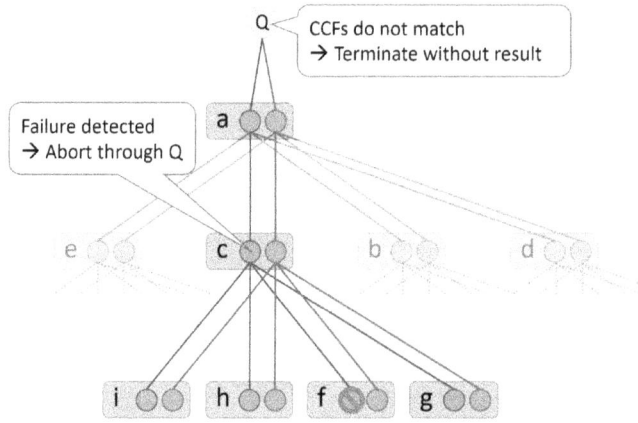

Fig. 2. *LowCost* strategy

Objective

LowCost is a radically optimistic strategy: it bets on an "almost ideal" world where little to no dropouts ever occur. It leverages the straw-man protocol explained in Algorithm 1, correcting its main issues (unchecked validity and lack of termination) with low-cost mechanisms to make it reliable. By saving on checks and verifications, the strategy is able to effectively minimize costs.

Design Principles

We observe that the straw-man protocol is near-optimal since (i) the data is sent up the tree only once by each node, and (ii) there is no sync between the parallel aggregation trees. The idea is to keep these desirable properties while still ensuring the protocol's validity and termination.

Protocol

Recall that contributors transmit their s shares simultaneously to the s leaf aggregators. In case of a contributor dropout, it is unlikely, but not impossible, that the shares are transmitted completely to, e.g., $s-1$ aggregators and incompletely to the last one.

- **Node replacement policy.** An aggregator is replaced only if its dropout occurs before the node receives any data from any of its children. For instance, node c in Fig. 2 can be replaced only if it drops out before receiving any data from i, f, h, or g. Replacing c after this would violate the 'send-only-once' principle since one or several of its children would need to re-send data.

- **Validity** It is ensured by leveraging the low-cost CCF mechanism which does not require any inter-tree communication, unlike any form of *sync*.
- **Termination.** *LowCost* terminates either (i) gracefully after the querier receives all s shares with consistent footprints from its children or (ii) abruptly if a child is asked to resend data by its replaced parent aggregator. Abrupt termination is done by having replacement node alert the querier, and they then both propagate the alert for other nodes to stop. Note also that due to the 'send-only-once' principle, any node can safely leave the protocol after it has sent its *IntRes* to its parent (i.e., a progressive termination from the leaves to the root).

Discussion

This basic protocol minimizes cost and is reliable. However, it has a binary behavior w.r.t. completeness. That is, completeness drops to 0% if (i) a single fatal aggregator dropout occurs or (ii) a single contributor drops out after sending only a sub-set of its s shares (thus leading to different versions in the parallel subtrees). This makes the completeness of *LowCost* extremely sensitive to dropouts. The reason is that there is no inter-tree sync mechanism allowing a convergence between the s parallel aggregates.

6.2 *HighCpl*

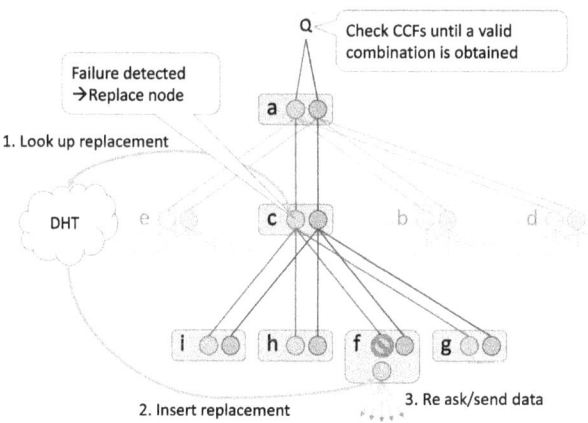

Fig. 3. *HighCpl* strategy

Objective

To have a complete view of the design space, we want to design a strategy that eagerly seeks to maximize completeness regardless of cost. We call this strategy *HighCpl*. Despite its more pessimistic outlook on node dropouts compared to

LowCost, *HighCpl* is designed to be optimistic about its ability to handle them, hence the unlimited efforts to detect and manage dropouts.

Design Principles

To achieve this objective *HighCpl* adopts the opposite behavior compared with *LowCost*: (i) any tree node (contributor or aggregator) can re-send its data whenever required (e.g., following a node replacement) and (ii) *HighCpl* propagates the data upward in the tree as fast as possible to maximize the chances of diffusion and then uses non-blocking *sync* for convergence between trees with eventual consistency thanks to CCF.

Protocol

- **Node replacement policy.** In *HighCpl*, any dropped out aggregator is replaced as soon as the dropout is detected by its parent node regardless if the dropout node has already sent one or several times its data up the tree, as illustrated in Fig. 3. The replacement node asks *IntRes* from its children after replacement and sends the aggregate to its parent if that aggregate has a different version (CF) compared with the last sent aggregate (recorded by the parent). This may happen, for instance with the replacement of a leaf aggregator with some dropped out contributors.
- **Validity.** Non-blocking sync is required at the leaf aggregator level to ensure validity despite contributor dropouts (a new version after a leaf aggregator replacement would anyway require a new synchronization). For the other levels, no sync is required since aggregators are replaced in case of dropouts and any new version sent at the leaf aggregator level triggers new computations up to the tree root.
- **Termination.** *HighCpl* can terminate after the querier receives s consistent shares from its children.

Discussion

HighCpl searches to maximize completeness through systematic dropout replacements, subsequent data re-sends, and minimalist non-blocking sync. The consequence is obviously an increased overhead since the same data can be transmitted and aggregated multiple times but in favorable cases, it should lead to reduced latency and maximized completeness.

6.3 *Sync&Prune*

Objective

The extreme behavior of *LowCost* and *HighCpl* may make them impractical to use in real-case scenarios. *Sync&Prune* aims to offer an adapted trade-off between completeness and cost, trying to minimize overhead without completely hurting completeness.

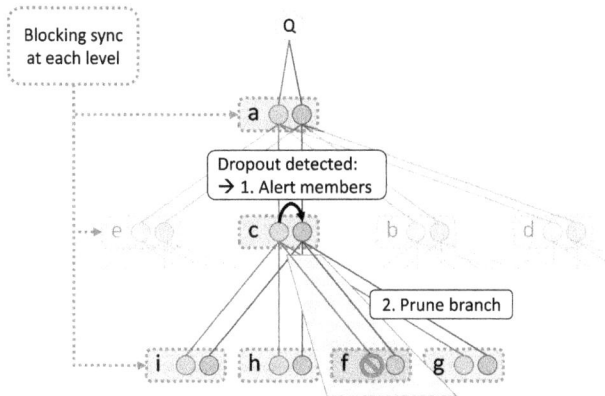

Fig. 4. *Sync&Prune* strategy

Design Principles

Sync&Prune is based on the *LowCost* strategy. It leverages the same 'send-only-once' principle to minimize cost like in *LowCost*. However, different from *LowCost*, *Sync&Prune* allows for convergence between the parallel trees by using *sync* and pruning the branches containing inconsistent shares or dropouts, as shown in Fig. 4.

Protocol

- **Node replacement policy.** As for *LowCost*, given the 'send-only-once' principle, nodes are only replaced as long as their children never sent data.
- **Validity.** Non-blocking sync is not compatible with the 'send-only-once' principle because if nodes are to send data only once, they should take a bit of time to make sure they are sending valid data. Thus *Sync&Prune* employs *blocking sync* to ensure validity and does so at every level of the tree. Indeed, since dropout aggregators are not replaced, synchronization is required at all levels to ensure a consistent result between the parallel trees. Syncing at each tree level allows *Sync&Prune* to progressively prune the tree branches corresponding to dropped out aggregators and branches producing invalid results.
- **Termination.** *Sync&Prune* terminates when the querier receives s shares from its child group or if it detects a dropout in the root group. Like in *LowCost*, lower-level nodes progressively terminate, from leaves to the root, after sending their *IntRes* to their parents. Whenever a branch of the tree is pruned, nodes of that branch are contacted using the remaining tree structure to inform them that they can terminate.

Discussion

Sync&Prune is expected to have a low cost due to the 'send-only-once' strategy. Syncing also adds an overhead, especially with the blocking variant, but we expect it to be low compared with the data transmission and the message decryption/encryption costs. The blocking sync is also expected to increase latency in comparison with *LowCost*. However, sync should fix the 0% completeness issue of *LowCost* by syncing and pruning sub-trees in case of dropouts, rather than aborting the whole execution.

6.4 *Hybrid*

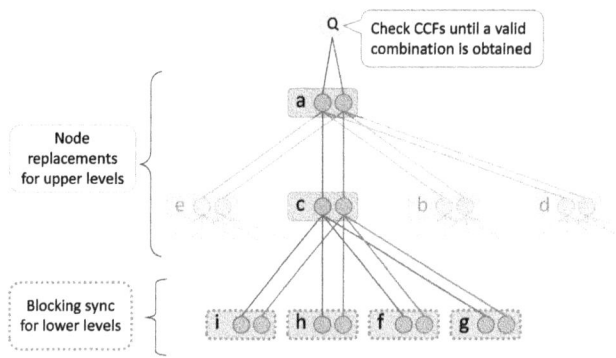

Fig. 5. *Hybrid* strategy

Objective

Hybrid is the second trade-off strategy attempting to maximize completeness and maintain a reasonable cost: by relaxing *HighCpl*'s dropout intolerance, *Hybrid* should avoid overpaying to replace nodes.

Design Principles

The idea is to have a hybrid approach combining principles from *HighCpl* and *Sync&Prune*. In *Hybrid*, the contributors employ a 'send-only-once' strategy like in *Sync&Prune*, while the aggregators re-send data whenever necessary like in *HighCpl*. The rationale is: (i) a significant part of the query cost comes from the data transmission at the contributors' level given a large number of contributors; and (ii) in the upper part of the tree, replacement induces less costs and provides comparatively more benefits, in terms of completeness.

Protocol

- **Node replacement policy.** In *Hybrid*, a leaf aggregator that drops out is replaced only if the dropout occurs before the node receives any data from its contributors to comply with the 'send-only-once' principle for contributors. On the other hand, any dropout aggregator in the upper levels is systematically replaced and requires their children to re-send their *IntRes*. Note that

when leaf aggregators drop out before they send their intermediate results, because of the 'send-only-once' policy, will prevent replacements from reproducing the intermediate results. In this case, the adopted behavior is to fall back on the *Sync&Prune* strategy of telling parent to prune the part of the tree containing the leaf group, as illustrated in Fig. 5.
- **Validity.** To minimize overhead, *Hybrid* uses a *blocking sync* strategy at the leaf aggregators. Thus, leaf aggregators send a single version of *IntRes* to their parent. Moreover, since leaf aggregators are not replaced if they drop out after receiving data from contributors, the sync at their parent's level induces pruning as in *Sync&Prune* , which negatively impacts completeness. However, the aggregator replacements in the upper levels avoid pruning a large number of contributors and thus favor completeness. Also, replacements in the upper levels can generate multiple versions.
- **Termination.** As in *HighCpl*, *Hybrid* can terminate after the querier receives s consistent *IntRes* from its child group. The contributor nodes and leaf aggregators nodes terminate after sending all the s shares to their parent(s).

Discussion

This hybrid strategy inherits the best of the two worlds: maximized completeness through replacement in the upper levels and limited cost due to 'send-only-once' at the contributors' level. However, it also inherits, but at a smaller scale, the limitations of the two approaches: some loss in completeness because of the non-replacement of leaf aggregators as well as some overhead generated by data re-sends in the upper levels.

Although the protocol of this strategy applies the *Sync&Prune* for leaves and contributors and *HighCpl* for upper aggregators, the depth threshold between those strategies could be adjusted. It is done automatically and at the branch level by groups being pruned out of the tree, but it can also be enforced at the protocol level by having contributors and any levels of aggregators above follow the *Sync&Prune* strategy, with extreme values of this threshold devolving either into *Sync&Prune* when the threshold in at a minimum, or into *HighCpl* when the threshold is at its maximum. However, the replacements have the biggest impact on performances for leaf aggregators, because that is where there is the most nodes to replace and where the most data is exchanged. Therefore, setting the threshold just above the leaf aggregators yields the best result, with higher threshold having only marginal benefits while slightly decreasing performances.

As mentioned for the node replacement policy, leaf aggregator groups can be pruned when some members send data while others are being replaced. This could lead to extreme cases where the parent group of a leaf group has almost no children left alive. A new security issue arises when only one child group is left: an attacker learns more than expected due to its nodes being present at adjacent depths of the tree and different group positions but observing the same data. This can be mitigated by limiting the number of replacements a parent group can do before being pruned.

Table 1. Design and building blocks

Strategy	send once	use sync	blocking sync	use CCF	replace aggregators
LowCost	yes	no	n/a	yes	no
HighCpl	no	leaf aggregators	no	yes	yes
Sync&Prune	yes	all levels	yes	no	no
Hybrid	contributors	leaf aggregators	yes	yes	yes

6.5 Summary

Table 1 summarizes the design and the behavior of each strategy. By eagerly replacing nodes, *HighCpl* favors completeness at the expense of cost while *LowCost* does not make any significant effort to favor completeness. *Sync&Prune* still favors low overhead but avoids the binary behavior of *LowCost* by pruning sub-trees. Finally *Hybrid* leverages the best of both *Sync&Prune* and *HighCpl*.

7 Performance Evaluation

In this section, we present the evaluation platform and used metrics, then describe the experimental parameters and the system security configuration. We present and analyze the experimental results varying the dropout rate and the other parameters, and then conclude with an analysis of the best-fitted strategy depending on the context.

7.1 Experimental Platform

Our main goal is to evaluate the four proposed strategies in a large P2P PDMS system wherein the nodes are structured leveraging a Chord DHT overlay [61]. However, peer-to-peer networks are notoriously hard to study: they rely on many participants across various geographical regions, using different physical networks and a wide range of hardware [7,23,44]. The most notable problems experiments on peer-to-peer networks are facing are:

– **Reproducibility**. The large number of moving pieces that compose a peer-to-peer network makes experimenting hard to reproduce. Good simulators enable reproducible results using deterministic randomness. Moreover, they should offer a way to perform fair ablation studies by keeping all things equal when only varying a parameter.
– **Scalability**. The study of peer-to-peer networks often relies on having thousands of nodes connected. Each node may need to independently simulate some local properties on top of doing any computations required by the protocol or application built on the network. Managing all these operations can be challenging at scale.

– **Application-specificity**. Having a detailed simulation of every network mechanism is not of equal importance depending on what is being tested: while it makes sense for new or modified versions of existing components, it may not always be true when testing higher-level applications built on peer-to-peer networks. Working at the appropriate level can greatly enhance scalability and reproducibility.

Existing Simulators

Intending to test the strategies we proposed in Sect. 6, we follow the same general approach as in the related work on P2P systems [31,50,61,71], i.e., our results are based on a simulator [18] which creates a logical DHT between simulated nodes. Indeed, simulators were used to evaluate the performance of the state-of-the-art structured DHTs (such as Chord [61] and CAN [50]) and systems that leverage P2P DHTs, such as in the distributed information retrieval area [31,51,63,64,71].

State-of-the-art peer-to-peer network simulators [44] focus either on scalability or fidelity. P2PSim [18] provides discrete-events simulation in a peer-to-peer network while offering several options to model the underlying physical network connecting peers. However, the amount of events that this application generates makes it hardly usable above a few thousand nodes. In PeerSim [43], up to hundreds of thousands of nodes can be simulated, however with a much lower fidelity since it only simulates query-cycles (from the moment a node queries some data in the network to the moment it receives it). In both cases, building more complex protocols on top of that simulator is ill-advised, because their focus is too specific to peer-to-peer communications rather than peer-to-peer and decentralized applications.

With the recent popularity of federated learning, the amount of tooling available to monitor and experiment on those systems has grown. The goal of those simulators is generally to experiment with new components of some FL workflow or to deploy them to production. FLINT [67] offers a platform that can provide many useful analytics and tools for users trying to transition from a centralized ML setting to a federated learning setting, as well as a decision framework to understand how worthy it can be to switch. However, it is focused on the cloud-based FL, not suitable for peer-to-peer usage, and favors efficiency over fairness by grouping devices by capabilities to accelerate training. More research-oriented platforms like Flower [10] on the contrary, focus on large-scale networks of heterogeneous devices, with up to 15 million devices simulated using only a pair of high-end GPUs. However, the dropouts model it considers resembles that of *SecAgg* [12], which is incompatible with our setting, as explained in Sect. 2.4.

Neither of the highly detailed or FL-focused simulators can however simultaneously address the complex dropout model that we are targeting in our strategies, the specific peer-to-peer communication overlay, and the application to large-scale FL that we require. This acts as our motivation to develop our simulator, tailored to our needs.

Simulator Design

The first step of designing a simulator is to define the scope. If it is too wide, the simulation gains in fidelity negatively impact scalability, and conversely. Our goal is to make the simulator as reusable as possible, to enable others to test and experiment with peer-to-peer federated learning on networks of PDMS. For this reason, we reduce the scope of the simulator to the core phase of our protocol, namely the aggregation phase. This allows experimenters using different components (such as constructing the tree overlay or disseminating models to contributors) to still benefit from the insights provided by our simulator while using their custom logic. This follows the approach used by framework like Flower [10] to focus on the aggregation while being compatible with many methods to organize the rest of the computations.

Given this scope, we can see that modeling communications at a low level is not necessary: once the tree overlay is fixed to transmit data, DHT routing is only used to find replacement nodes. Because of this limited impact, a probabilistic approach to find the number of hops needed to route messages between nodes can save a lot of computing during the simulation.

To make the simulation easy to analyze, we used an event-driven simulator: every action in the protocol is modeled as a message that is chronologically inserted in a queue. For variable duration interactions, a first message is used to indicate the start of the interaction and a second message that can be postponed indefinitely while the node is busy indicates the end. The queue can be stored on the hard drive for later analysis.

The simulator is designed with modularity in mind, as we need to test multiple building blocks and plan on having others experiment with new blocks. To this end, the simulator is conceived to be easily parameterized, and adding new components is straightforward. Building blocks can then be mixed and matched according to the experimental setting.

Finally, many statistics are collected at every step of the simulation. These statistics can either be aggregated when analyzing many executions of the protocol, or they can be maximally detailed when needed. The type of statistics collected can be adjusted for each experiment, to avoid exporting too much data if this is not needed.

7.2 Evaluation Metrics

Our experimental evaluation is focused on the tension between cost and completeness in the proposed protocols for different parameters impacting security and/or performance. With respect to cost, our simulator captures the typical metrics for evaluating distributed protocols. At the network level, we measure the required **bandwidth** (or bytes per query) at the node or system levels. The consumed bandwidth is of particular interest, especially in the FL context wherein transmitted model parameters can have a significant size (see Table 2). The required amount of **work** (or CPU time per query) at the node or system levels is equally important. Finally, we also need to estimate the **latency** to process a query. While estimating bandwidth is simple by monitoring the exchanged messages and the size of their content, estimating absolute time values (i.e., in seconds) for work and latency is more challenging since such measures are highly dependent on the context (underlying network topology, PDMS node heterogeneity, network congestion, etc.). For the network, we consider network links with latency and bandwidth based on average values of domestic Internet in France [46] and add random noise to these values to be more representative of PDMSs heterogeneous connections. For the local work on each PDMS impacting the total work and the latency, we consider the most costly operations during the protocols, i.e., the cryptographic operations. We note that in practice, no participant can have the exact value of completeness for a given result: although the number of contributions in the final aggregate is computed, the number of nodes that intended to contribute can never be known since some of them may drop out before telling anyone about their intentions.

To calibrate the simulator (see Table 2), we measured on a standard laptop computer equipped with Intel i5-9400H CPU @ 2.50GHz the cost of classical asymmetric encryption for signing and verifying messages, which is required to open the secure channels between communicating nodes. We also measured the time required by contributors/aggregators to process their data (e.g., encryption/decryption of a model of different sizes using AES256).

7.3 Experimental and Security Parameters

Besides the simulator calibration described above and summarized in purple in Table 2, the other experimental parameters are divided into four classes as follows.

- **System setup and security.** We consider a large P2P system of $N = 10^6$ nodes. We consider that the most powerful attacker can control up to C nodes. Our goal is to avoid data leakage with a very high probability (e.g., a value of the security threshold $\alpha < 10^{-6}$). To determine the number of shares s, we used the revised version of the formula presented in Sect. 5.2 and allowed only one replacement. For simplicity, we set s and deduce, depending on α and N, the maximum number of controlled nodes C. With $\alpha = 10^{-6}$, the group size of 4, 5, or 6 allows to tolerate up to, respectively, 21K, 44K, or 72K colluding nodes. All these values are quite high since it must be the same attacker that controls all these nodes (see also [13]).
- **Dropout rate and simulation of dropouts.** We vary the dropout rate from none up to extreme values, considering the most interesting, medium-range values. The no-dropout case allows providing a lower bound for the cost metrics. The extreme dropout rates are not representative of a real system setup (e.g., 1% dropout rate means that all the nodes drop out –for that query– after 100 s!) but allow observing trends and limitations of each strategy. Note also that dropouts during a query are only related to that query (i.e., nodes may be still working correctly, e.g., for the DHT overlay). Finally, we should stress that node dropouts are pre-computed before the query exe-

cution and independently of the strategy to produce the exact same dropouts at the same moment and allow a fair comparison of the different strategies.
– **System scalability.** We use a fan-out of 8 for the aggregation trees since this value offers the best trade-off between latency and total work in our setting (see [42] for the fan-out tuning detail). To measure the system scalability, we consider different values for selectivity and model size. The selectivity determines the number of contributors for a query and consequently, the aggregation tree height (the tree height of 3, 4 and 5 corresponds, with a fan-out of 8, to a selectivity of respectively 0.05%, 0.4% and 3.2%). For the model size, we considered very small (1KB) to large (16 MB) models to cover a wide range of FL applications. For instance, [12,59] consider the size of 1MB.
– **Number of runs and box plots.** Given the randomness of simulated dropouts, and despite the pre-computation of dropouts (see above), there is a lot of variability between simulation runs. Indeed, a single dropout can have dramatically different outcomes depending on where in the tree and when it occurs. We aggregate the results of 50 runs to obtain statistically representative results. In addition, we use box plots which help visualize this variability and the distribution of runs. In the figures, the lower and upper whiskers of the box respectively represent the min and the max for the plotted metric, and the lower and upper edges of the box represent the first and third quartile respectively. To make the boxes easier to read, we exclude outliers (i.e., points that are further away than 1.5 in the interquartile range) which are directly represented as points. Finally, the line connecting the mean values is also represented.

7.4 Performance with Varying Dropout Rate

Now that the experimental setup has been described, we present the performance of our strategies under specific scenarios, varying parameters one by one. We start by varying the dropout rate, as it is the most important parameter to understand how dropouts affect the system.

Table 2. Simulation parameters

Description (*notation*)	Values (*default*)
Network latency [46]	30ms
Network bandwidth [46]	6MB/s
Asymmetric cryptographic operation	10ms/op
Local processing including symmetric crypto	5ms/MB
Number of PDMS nodes (N)	10^6
Group size (s), based on α, N, C	4, 5 or 6 (5)
Maximum number of replacement per group	1
Percentage of node dropouts per second (D)	0% to 1.2% (0.25%)
Aggregation tree fanout (f)	8
Aggregation tree height (h)	3, 4, or 5 (4)
ML model size (M)	1KB to 16MB (1MB)
Number of runs (to account for variability)	50

Figures 6, 7, 8 and 9 depict respectively the completeness, latency, bandwidth and work by node, varying the dropout rate D for the four studied strategies. All strategies, when using the same parameters, perform almost identically when there are no dropouts since the overhead of each strategy (e.g., blocking sync) is mostly negligible compared to the transmission cost of the 1MB model through the tree. This setting provides a nice baseline to which we can then compare strategies in different settings.

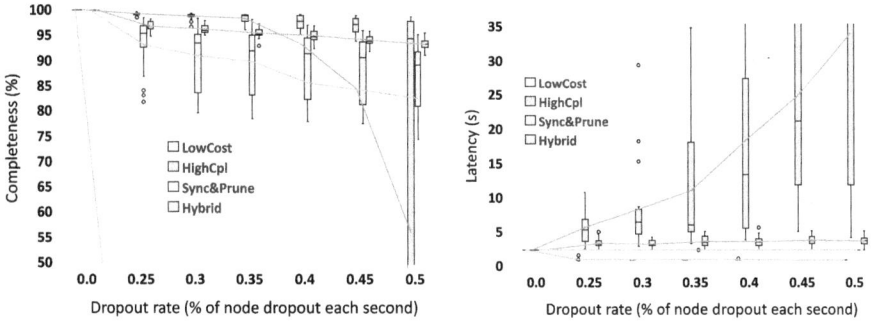

Fig. 6. Completeness, 1 MB model   **Fig. 7.** Latency, 1 MB model

LowCost Cannot Handle Even a Few Dropouts

As soon as some nodes drop out, *LowCost* fails to obtain any result. The strategy uses an abrupt termination (see Sect. 6.1), sometimes even before contributors can contribute data. Figure 8 shows that despite early termination, some contributors do manage to send their shares, which consumes some bandwidth even though these data are never used. In the rest of the study, *LowCost* is often ignored because we focus on settings where dropouts generally prevent it from succeeding. Specifically, under default parameters and few dropouts (not shown in the figures), we obtained 45% of *LowCost* successful executions with $D = 0.01\%$, only 10% when $D = 0.02$ and 0% with higher dropout rates.

HighCpl has the Best Completeness But Degrades Rapidly with Large Dropout Rates

HighCpl degrades rapidly when $D \geq 0.4$ with a variability between runs that explodes. Indeed, the overhead induced by dropout handling increases significantly with the dropout rate (see Figs. 7 and 8). *HighCpl* gives up replacing nodes after having exhausted the maximum number of replacement nodes. On the one hand, this allows maintaining a path between the remaining contributors and the root of the tree, which in turn allows for aggregates to move faster up the aggregation path. On the other hand, because data are eagerly sent along the parallel trees without explicit synchronization mechanisms, it can generate a lot of aggregate re-sends before the final CCF convergence, impacting work, latency, and bandwidth. Hence, *HighCpl* is subject to a snowball effect: during the time needed to replace some nodes, even more nodes drop out, preventing the querier from obtaining valid aggregates. This makes it unsuitable for large dropout rates.

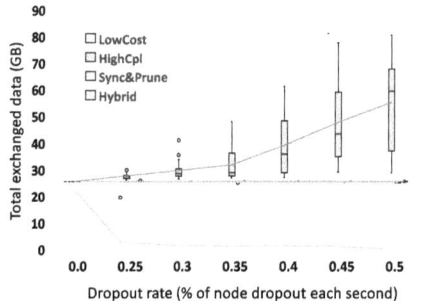

Fig. 8. Bandwidth, 1 MB model

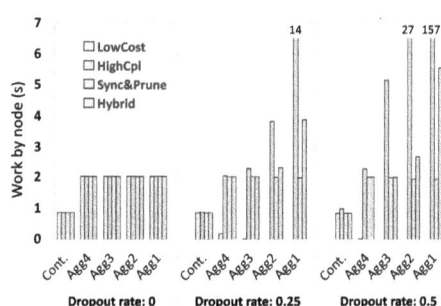

Fig. 9. Work by node, 1 MB model

Sync&Prune has Stable Costs but Lower Completeness

Opposite to *HighCpl*, *Sync&Prune* starts out with a lower completeness and higher dispersion. The rationale is that for *Sync&Prune*, completeness is determined by the location of dropouts in the tree, with dropouts higher in the tree severely affecting it. However, *Sync&Prune* has a more stable performance on all metrics as the dropout rate increases, because pruning trees is cheaper than resending data that may become invalid: it puts an upper bound on costs at the expense of initially lower completeness. Since the strategy ensures that any node works at most once, dropouts reduce the load of their parents, thus, this strategy is beneficial for cost, but has a negative impact on completeness.

Hybrid Combines the Best of the two Previous Strategies

Hybrid has stable completeness, work, and bandwidth throughout the dropout rate spectrum. Compared with *HighCpl*, the completeness is lower for low dropout rates because contributors never resend their data and instead, leaf groups and their contributors are pruned. It is however a lot better for high dropout rates because leaf aggregators are never replaced, preventing the conflicting CCFs that cause the snowball effect. To our surprise, *Hybrid* has only about $0.8 - 1.7\%$ larger communication cost than *Sync&Prune* but up to 20% higher latency. The rationale is that at the leaf aggregators level, where the vast majority of the protocol bandwidth is consumed, both protocols behave the same. In upper levels, *Hybrid* can resend data, inducing an increased latency, but this has a limited impact due to blocking **sync**.

HighCpl Suffers from Too Many Aggregate Versions

Figure 9 shows the distribution of the work across the tree layers, i.e., the average work per node in each tree level, for each strategy. We observe first that the work per aggregator in all levels is similar when the dropout rate is low. Thanks to our tree structure, the load is effectively and fairly distributed across system nodes. Contributors have less work to do in this setting since they transmit $k = 5$ shares while aggregators receive and process $f = 8$ shares and send one more to their parent. We also observe that the snowball effect in *HighCpl* mainly affects the aggregators closer to the root since those nodes receive the largest number of different aggregate versions. *Hybrid* also concentrates the load on higher level aggregators but manages to keep this overhead in a reasonable range thanks to the synchronization at the leaf aggregators.

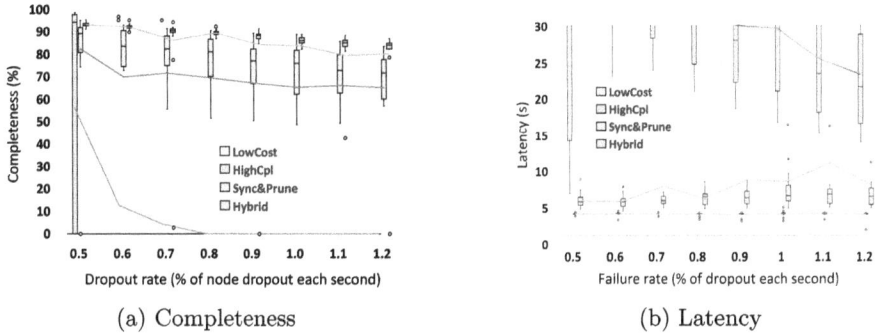

(a) Completeness (b) Latency

Fig. 10. Performances under extreme dropout rates

Hybrid and *Sync&Prune* Tolerate Extreme Dropout Rates

Figure 10a presents the behavior of our strategies with extreme dropout rates ($D \geq 0.5$). With dropout rates $D > 0.7$, *HighCpl* fails to obtain any result for most of its executions, but *Sync&Prune* and *Hybrid* only have a small reduction in completeness because they are pruning the nodes and branches that prevent the execution from finishing. However, costs for *Hybrid* grow faster than those of *Sync&Prune*, as shown in Fig. 10b. We note that despite wider boxes for *Sync&Prune*, *Hybrid* has some outlying executions (even with no results) that fail to complete. Indeed with extreme dropout rates, the maximum number of replacements can be reached leading to pruning entire sub-trees and thus, reducing drastically completeness. Reaching this limit also explains why *HighCpl*, which replaces more nodes, aborts the execution sooner as the dropout rate increases.

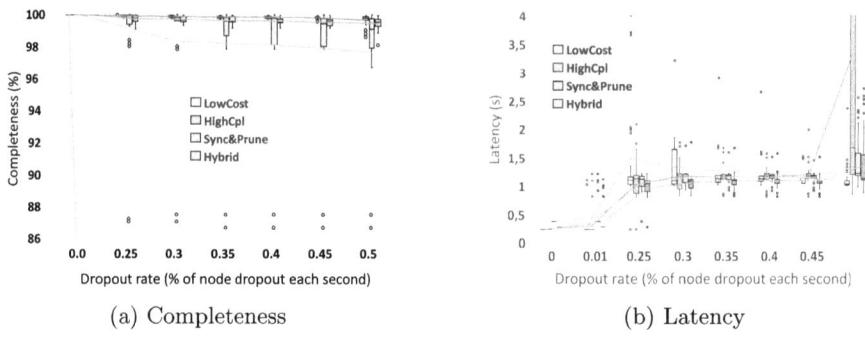

Fig. 11. Performances for 1KB model

HighCpl and *Hybrid* Work Well with Small Models

Figure 11a and Fig. 11b show the obtained completeness and latency with a small model of only 1KB. Small models change the protocol bottlenecks since the model transmission becomes less impacting. Moreover, since the overall latency is much smaller, there are fewer dropouts and thus better completeness for all strategies. *HighCpl* and *Hybrid* obtain almost 100% completeness, much better than *Sync&Prune*, with low overhead. Figure 11a shows also something interesting: we can observe a set of outlier executions for *Sync&Prune* with a completeness of about 87% (i.e., missing a fraction of 1/8 of the results). This is explained by the dropout of an aggregator in the second level group happening after its children have sent some data and are generally the reason for the higher variability of completeness in the executions of *Sync&Prune*. We can finally see in Fig. 11b that latency sharply increases as soon as dropouts start to happen, even for the smallest dropout rates where we can see a pack of outlier at around 1 s latency. This occurs because dropouts are detected after a timeout, whose duration is significant compared to the overall latency: active nodes will consider a peer as dropped out if it did not respond to the most recent health check in up to 10 times the average duration of a health check, which in this case should take ≈ 60ms. This leads to a 600 ms delay, which can be observed by the difference of latency between runs without dropouts and runs with higher dropout rates.

7.5 Scalability and Security

Now that we have extensively studied the robustness of our strategies, we want to focus on the scalability and security aspects, as they are key issues for decentralized federated learning.

Very Large Models Require Less Dropouts

For a model size of 16 MB, none of the strategies reaches completeness above 50% as shown in Fig. 12a. We can see that *Sync&Prune* starts outperforming *Hybrid*, which only manages to finish a few low completeness (10%) executions. This

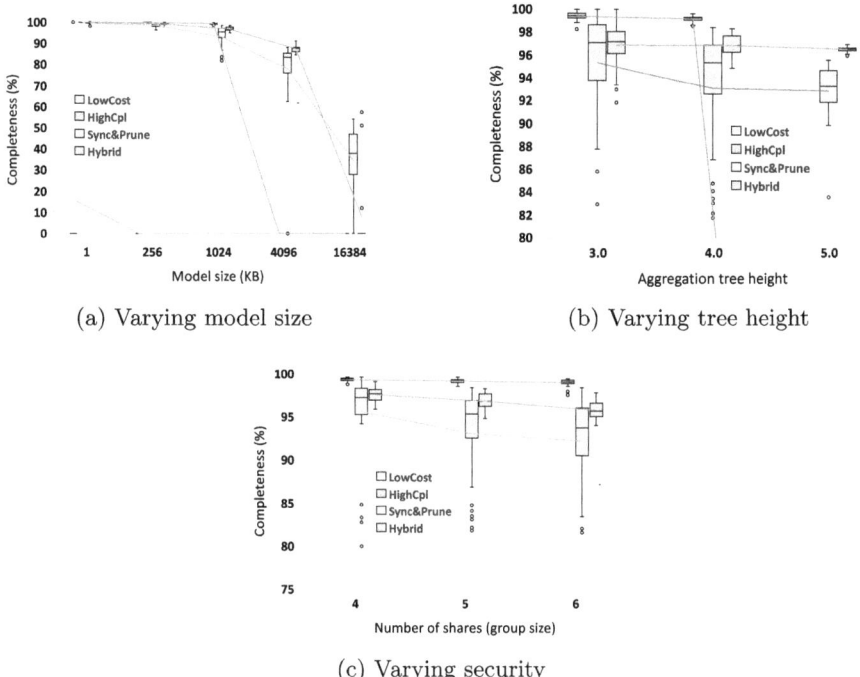

Fig. 12. Completeness for different parameters variations

is linked to the snowball effect caused by aggregator replacements: even if most nodes are not replaced (i.e., the leaf aggregators), the replacements higher in the tree are enough to create conflicts in the aggregate versions, which cannot be resolved since models transmission takes time, thereby leading to more dropouts. *Sync&Prune* is outperforming here because it only sends once the data that has been explicitly agreed upon by all group members, and prunes branches that fail to come to an agreement. More generally, we can conclude that very long queries (due to the transmission of the 16 MB model) are not compatible with the default dropout rate (with too many dropouts) while 4 MB models are correctly supported.

Hybrid and *Sync&Prune* Support Large Aggregation Trees

Another important aspect of scalability is the number of participants contributing to the query, which is characterized by the selectivity σ in our system. We use the tree height as a proxy for the number of contributors because it has more impact on performances: the selectivity in a given network can vary almost without affecting performances until the tree is saturated and the depth increases, where we observe sharp changes in metrics. Using the tree height prevents gaps in the plots due to the effect of rounding the depth and reduces the complexity of the analysis. *HighCpl* is once again failing when the height is greater than 4, taken down by the snowball effect of 8 times more nodes constantly resending new aggregate versions and never converging to a stable version. *Sync&Prune* and *Hybrid* on the other hand are scalable with an increasing number of contributors showing that they can fully benefit from the distributed nature of the execution. The seemingly increased stability comes from the fact that some runs being further away from the performances of most runs, which labels them as outliers and exclude them from the box and puts a point instead. The fact that it barely affects boxes demonstrates the small impact of large trees on completeness, where failures happen most where they have the least impact: on leaves.

Larger Groups (Better Security) are Well Supported

Finally, we study the impact of the security parameter on all strategies. We can see in Fig. 12c that increasing the security only has a mild impact on completeness for *Sync&Prune* and *Hybrid* while *HighCpl* remains mostly unaffected. The opposite is however happening in terms of work per node. Increasing the group size makes synchronization harder for *Sync&Prune* and *Hybrid*, resulting in more frequent pruning. The security parameter also impacts *HighCpl* by requiring more work to make the parallel trees come to an agreement.

7.6 Best Fit Strategy

		LowCost	HighCpl	Sync&Prune	Hybrid	
Height	Size	0	0.01	0.25	0.5	1
3	1 K	100%	100%	100%	100%	100%
4	1 K	100%	100%	100%	100%	99%
3	1 M	100%	100%	99%	99%	96%
4	1 M	100%	100%	99%	93%	84%
3	4 M	100%	100%	97%	78%	59%
4	4 M	100%	100%	87%	68%	28%

CRITERIA: Best completeness

(a) At max completeness

		LowCost	HighCpl	Sync&Prune	Hybrid	
Height	Size	0	0.01	0.25	0.5	1
3	1 K	100%	100%	100%	99%	97%
4	1 K	100%	95%	99%	98%	91%
3	1 M	100%	80%	89%	90%	87%
4	1 M	100%	100%	93%	83%	84%
3	4 M	100%	100%	81%	60%	59%
4	4 M	100%	100%	78%	51%	28%

CRITERIA: Least total work, at least 80% of best completeness

(b) At 80% completeness

		LowCost	HighCpl	Sync&Prune	Hybrid	
Height	Size	0	0.01	0.25	0.5	1
3	1 K	100%	100%	100%	99%	97%
4	1 K	100%	95%	99%	98%	99%
3	1 M	100%	100%	97%	99%	96%
4	1 M	100%	100%	97%	93%	84%
3	4 M	100%	100%	97%	76%	59%
4	4 M	100%	100%	87%	68%	28%

CRITERIA: Least total work, at least 95% of best completeness

(c) At 95% completeness

Fig. 13. Least costly strategy for various completeness threshold

In this last section, we first choose some typical parameter settings that best exemplify realistic scenarios, then define different objectives that guide us in comparing the strategies. Finally, we report in Fig. 13a, 13b and 13c the region of the parameter space where each strategy performs best. We study two heights for the trees, corresponding to aggregating data from 500 and 4000 contributors. It includes the default height as well as a smaller height since they are representative of the context of federated learning (i.e., it is pertinent to aggregate from fewer contributors more frequently). We choose three model sizes: one for tiny models (e.g., for training simple models or computing aggregate statistics), and the default value of 1MB and then 4MB (which correspond to models used in a wider range of applications, ranging from image classification to natural language processing). The last parameter that we vary is the dropout rate to observe the robustness of each strategy.

Our first objective is simply the highest completeness. Since a marginal increase in completeness can be costly, and unfairly advantage a strategy, we select, for the second and third objective, the strategy that incurs the least work after a pre-selection of strategies reaching a completeness of 80% and 95% of the strategy with the best completeness. We present the results in Fig. 13a, 13b and 13c in the form of a table where rows correspond to a couple of height and model size and columns to a dropout rate. The color of each cell corresponds to the strategy that best fits the given objective while the value is the averaged completeness of 50 executions.

***LowCost* Best Fits with Very Few or no Dropouts.** Without dropouts, *LowCost* always wins. All strategies have 100% completeness in this case but *LowCost* has the lowest cost since parallel trees are not synchronized. The same happens with very few dropouts and small models.

***HighCpl* Maximizes Completeness but is Costly.** In Fig. 13a, *HighCpl* is the strategy that offers the best completeness in a majority of cases, i.e., except for bigger models or extreme dropout rates. It is also the best strategy in a few settings on Fig. 13c.

***Sync&Prune* is Efficient and Reaches 80% Best Completeness.** When being more tolerant about completeness, *Sync&Prune* often becomes preferable because of its lower cost. With 80%, it even outperforms in almost all cases. *Hybrid* performs better for extreme dropout rates because maintaining the tree prevents pruning the most impacting aggregators. With 95%, *Sync&Prune* can still outperform in some cases (Fig. 13c), although in those cases, all strategies perform well anyway or oppositely, *Sync&Prune* is the only strategy that gets a result.

***Hybrid* is Efficient and Reaches 95% Best Completeness.** Hybrid fulfills its role as a trade-off strategy for every objective. It mostly outperforms in settings where *HighCpl* suffers from the snowball effect and *Sync&Prune* already has degraded performances. Moreover, for high completeness (i.e., the 95% objective), *HighCpl* has a smaller overhead compared with *Sync&Prune* .

8 Conclusion

Personal Data Management Systems arrive at a rapid pace allowing users to share their personal data within large P2P communities, which opens exciting perspectives. Federated learning is a prime example that could benefit from this abundant, diverse and complete source of personal data to train high quality ML models. However, this requires new protocols that protect the users' privacy and are adapted to the fully-decentralized nature of the PDMS ecosystem. To this end, we proposed a set of secure aggregation protocols for federated learning which are fully-decentralized, scalable, accurate and reliable. We analyzed the secure aggregation problem in the P2P PDMS context and showed that reliability is a key aspect raising tension between the potential completeness of the result and the aggregation cost. We then proposed four protocols having different trade-offs between completeness and cost. We extensively evaluated these protocols for a wide range of settings of the dropout rates, security settings, trained model size, or contributors' selectivity. Our results showed that these protocols can offer high completeness results at a reasonable cost in a wide range of settings.

We plan to extend this work in several directions and shortly discuss two of them. First, our current aggregation protocols consider a honest-but-curious threat model. Although this model is widely used in the secure aggregation context, it would be useful to consider more advanced attack models, wherein the attacker can completely control some system nodes. However, this has profound implications such as attacks on the DHT, on the tree construction mechanism and on the assumed uniformly random selection of nodes to create groups as well as the node replacement mechanism. We believe that the techniques proposed in [13], namely collaborative probabilistic proofs could probably be applied even if node dropouts make the problem more challenging. Second, the aggregation process can be prone to data poisoning and backdoor attacks [6,68]. The typical solution based on filtering malicious contributions cannot work with secure aggregation since contributors' inputs are private. Hence, rightful inputs cannot be distinguished from purposefully crafted data intended to corrupt a model training. Existing work [17] provides privacy-preserving ways to ensure data fits a criterion but using a centralized architecture. By supposing that PDMS data is signed by the entity that inserts it, we envision including these signatures into our CCF mechanism and aggregating them to allow participants to verify that the aggregate does not contain maliciously crafted data.

Acknowledgments. This work has been supported by the ANR 22-PECY-0002 IPOP (Interdisciplinary Project on Privacy) project of the Cybersecurity PEPR.

References

1. Agarwal, N., Suresh, A.T., Yu, F.X., Kumar, S., et al.: cpSGD: communication-efficient and differentially-private distributed SGD. In: NeurIPS (2018)
2. Alvim, M.S., Chatzikokolakis, K., Palamidessi, C., Pazii, A.: Local differential privacy on metric spaces: optimizing the trade-off with utility. In: IEEE CSF (2018)
3. Anciaux, N., Bonnet, P., Bouganim, L., Nguyen, B., et al.: Personal data management systems: the security and functionality standpoint. Inf. Syst. **80**, 13–35 (2019)

4. Anciaux, N., Bouganim, L., Pucheral, P., Sandu-Popa, I., et al.: Personal database security and trusted execution environments: a tutorial at the crossroads. PVLDB (2019)
5. ARM: Building a Secure System using TrustZone Technology (2008)
6. Bagdasaryan, E., Veit, A., Hua, Y., Estrin, D., et al.: How to backdoor federated learning. In: International Conference on Artificial Intelligence and Statistics, pp. 2938–2948 (2020)
7. Basu, A., Fleming, S., Stanier, J., Naicken, S., et al.: The state of peer-to-peer network simulators. ACM Comput. Surv. (CSUR) **45**(4), 1–25 (2013)
8. Bater, J., Elliott, G., Eggen, C., Goel, S., et al.: SMCQL: secure query processing for private data networks. PVLDB **10**(6), 673–684 (2017)
9. Bellet, A., Guerraoui, R., Taziki, M., Tommasi, M.: Personalized and private peer-to-peer machine learning. In: AIStat (2018)
10. Beutel, D.J., Topal, T., Mathur, A., Qiu, X., et al.: Flower: a friendly federated learning research framework. arXiv preprint arXiv:2007.14390 (2020)
11. Bittau, A., Erlingsson, Ú., Maniatis, P., Mironov, I., et al.: Prochlo: strong privacy for analytics in the crowd. In: Proceedings of the 26th Symposium on Operating Systems Principles (2017)
12. Bonawitz, K., Ivanov, V., Kreuter, B., Marcedone, A., et al.: Practical secure aggregation for privacy-preserving machine learning. In: ACM CCS (2017)
13. Bouganim, L., Loudet, J., Sandu Popa, I.: Highly distributed and privacy-preserving queries on personal data management systems. VLDB J. **32**(2), 415–445 (2023)
14. Brasser, F., Müller, U., Dmitrienko, A., Kostiainen, K., et al.: Software grand exposure: SGX cache attacks are practical. In: 11th USENIX Workshop on Offensive Technologies (2017)
15. Bulck, J.V., Minkin, M., Weisse, O., Genkin, D., et al.: Foreshadow: extracting the keys to the intel SGX kingdom with transient out-of-order execution. In: 27th USENIX Security Symposium (2018)
16. Cormode, G., Kulkarni, T., Srivastava, D.: Answering range queries under local differential privacy. PVLDB (2019)
17. Corrigan-Gibbs, H., Boneh, D.: Prio: private, robust, and scalable computation of aggregate statistics. In: NSDI (2017)
18. MISC
19. Cozy Cloud: Cozy Cloud (2023). https://cozy.io/fr/
20. Ding, B., Kulkarni, J., Yekhanin, S.: Collecting telemetry data privately. In: Advances in Neural Information Processing Systems (2017)
21. Dong, Y., Chen, X., Li, K., Wang, D., et al.: FLOD: oblivious defender for private byzantine-robust federated learning with dishonest-majority. In: ESORICS (2021)
22. Dwork, C.: Differential privacy. In: Automata, Languages and Programming (2006)
23. Ebrahim, M., Khan, S., Hasan Mohani, S.S.U.: Peer-to-peer network simulators: an analytical review. Asian J. Eng. Sci. Technol. (2012)
24. Erlingsson, Ú., Pihur, V., Korolova, A.: RAPPOR: randomized aggregatable privacy-preserving ordinal response. In: Proceedings of the 2014 ACM SIGSAC Conference on Computer and Communications Security, CCS 2014 (2014)
25. Eskandarian, S., Zaharia, M.: ObliDB: oblivious query processing for secure databases. Proc. VLDB Endow. (2019)
26. Fereidooni, H., Marchal, S., Miettinen, M., Mirhoseini, A., et al.: SAFELearn: secure aggregation for private federated learning. In: IEEE SPW (2021)

27. Ghazi, B., Manurangsi, P., Pagh, R., Velingker, A.: Private aggregation from fewer anonymous messages. In: Canteaut, A., Ishai, Y. (eds.) EUROCRYPT 2020. LNCS, vol. 12106, pp. 798–827. Springer, Cham (2020). https://doi.org/10.1007/978-3-030-45724-2_27
28. Gueron, S.: Memory encryption for general-purpose processors. IEEE Secur. Priv. **14**, 54–62 (2016)
29. Guo, X., Liu, Z., Li, J., Gao, J., et al.: VeriFL: communication-efficient and fast verifiable aggregation for federated learning. IEEE Trans. Inf. Forensics Secur. (2021)
30. Gupta, P., Li, Y., Mehrotra, S., Panwar, N., et al.: Obscure: information-theoretic oblivious and verifiable aggregation queries. PVLDB **12**(9), 1030–1043 (2019)
31. Hayek, R., Raschia, G., Valduriez, P., Mouaddib, N.: Summary management in P2P systems. In: EDBT (2008)
32. Huba, D., Nguyen, J., Malik, K., Zhu, R., et al.: Papaya: practical, private, and scalable federated learning. Proc. Mach. Learn. Syst. **4**, 814–832 (2022)
33. Kasiviswanathan, S.P., Lee, H.K., Nissim, K., Raskhodnikova, S., et al.: What can we learn privately? SIAM J. Comput. **40**(3), 793–826 (2011)
34. Li, N., Li, T., Venkatasubramanian, S.: t closeness: privacy beyond k-anonymity and l-diversity. In: ICDE (2007)
35. Loudet, J., Sandu-Popa, I., Bouganim, L.: DISPERS: securing highly distributed queries on personal data management systems. PVLDB (2019)
36. Loudet, J., Sandu-Popa, I., Bouganim, L.: SEP2P: secure and efficient P2P personal data processing. In: EDBT (2019)
37. Machanavajjhala, A., Kifer, D., Gehrke, J., Venkitasubramaniam, M.: L-diversity: privacy beyond k-anonymity. ACM Trans. Knowl. Discov. Data (2007)
38. Mansouri, M., Önen, M., Jaballah, W.B., Conti, M.: SoK: secure aggregation based on cryptographic schemes for federated learning. PETS (2023)
39. McMahan, B., Moore, E., Ramage, D., Hampson, S., et al.: Communication-efficient learning of deep networks from decentralized data. In: PMLR (2017)
40. Merkle, R.C.: A digital signature based on a conventional encryption function. In: Pomerance, C. (ed.) CRYPTO 1987. LNCS, vol. 293, pp. 369–378. Springer, Heidelberg (1988). https://doi.org/10.1007/3-540-48184-2_32
41. Mirval, J., Bouganim, L., Sandu Popa, I.: Federated learning on personal data management systems: decentralized and reliable secure aggregation protocols. In: SSDBM (2023)
42. Mirval, J., Bouganim, L., Sandu-Popa, I.: Practical fully-decentralized secure aggregation for personal data management systems. In: SSDBM (2021)
43. Montresor, A., Jelasity, M.: PeerSim: a scalable P2P simulator. In: P2P (2009)
44. Naicken, S., Livingston, B., Basu, A., Rodhetbhai, S., et al.: The state of peer-to-peer simulators and simulations. ACM SIGCOMM **37**(2), 95–98 (2007)
45. Nguyen, J., Malik, K., Zhan, H., Yousefpour, A., et al.: Federated learning with buffered asynchronous aggregation. In: AISTATS (2022)
46. nPerf: Baromètre des connexions Internet fixes en France métropolitaine (2020). https://perma.cc/DP8V-5ABT
47. Oleksenko, O., Trach, B., Krahn, R., Silberstein, M., et al.: Varys: protecting SGX enclaves from practical side-channel attacks. In: USENIX ATC, Boston, MA (2018)
48. Pilet, A.B., Frey, D., Taıani, F.: Robust privacy-preserving gossip averaging. In: SSS (2019)
49. Przydatek, B., Song, D., Perrig, A.: SIA: secure information aggregation in sensor networks. In: SenSys (2003)

50. Ratnasamy, S., Francis, P., Handley, M., Karp, R.M., et al.: A scalable content-addressable network. In: ACM SIGCOMM (2001)
51. Reynolds, P., Vahdat, A.: Efficient peer-to-peer keyword searching. In: Middleware (2003)
52. Rivest, R.L., Adleman, L., Dertouzos, M.L.: On data banks and privacy homomorphisms. Found. Secure Comput. **4**(11), 169–180 (1978)
53. Sabater, C., Bellet, A., Ramon, J.: An accurate, scalable and verifiable protocol for federated differentially private averaging. JMLR (2022)
54. Sabt, M., Achemlal, M., Bouabdallah, A.: Trusted execution environment: what it is, and what it is not. In: Trustcom (2015)
55. Sandu-Popa, I., Ton-That, D.H., Zeitouni, K., Borcea, C.: Mobile participatory sensing with strong privacy guarantees using secure probes. GeoInformatica **25**, 533–580 (2021)
56. van Schaik, S., Kwong, A., Genkin, D., Yarom, Y.: SGAxe: how SGX fails in practice (2020)
57. van Schaik, S., Minkin, M., Kwong, A., Genkin, D., et al.: CacheOut: leaking data on intel CPUs via cache evictions. In: SP (2021)
58. Schwarz, M., Weiser, S., Gruss, D., Maurice, C., et al.: Malware guard extension: using SGX to conceal cache attacks. In: DIMVA (2017)
59. So, J., Güler, B., Avestimehr, A.S.: Turbo-aggregate: breaking the quadratic aggregation barrier in secure federated learning. JSAIT **2**(1), 479–489 (2021)
60. So, J., He, C., Yang, C.-S., Li, S., et al.: Lightsecagg: a lightweight and versatile design for secure aggregation in federated learning. MLSys **4**, 694–720 (2022)
61. Stoica, I., Morris, R., Karger, D., Kaashoek, M.F., et al.: Chord: a scalable peer-to-peer lookup service for internet applications. ACM SIGCOMM **31**, 149–160 (2001)
62. Sweeney, L.: k-anonymity: a model for protecting privacy. Int. J. Uncertain. Fuzziness Knowl.-Based Syst. (2002)
63. Tang, C., Dwarkadas, S.: Hybrid global-local indexing for efficient peer-to-peer information retrieval. In: NSDI (2004)
64. Tang, C., Xu, Z., Dwarkadas, S.: Peer-to-peer information retrieval using self-organizing semantic overlay networks. In: ACM SIGCOMM (2003)
65. To, Q., Nguyen, B., Pucheral, P.: Private and scalable execution of SQL aggregates on a secure decentralized architecture. ACM TODS **41**(3), 1–43 (2016)
66. Triastcyn, A., Faltings, B.: Federated learning with Bayesian differential privacy. In: IEEE BigData (2019)
67. Wang, E., Chen, B., Chowdhury, M., Kannan, A., Liang, F.: FLINT: a platform for federated learning integration. Proc. Mach. Learn. Syst. **5**, 21–34 (2023)
68. Wang, H., Sreenivasan, K., Rajput, S., Vishwakarma, H., et al.: Attack of the tails: yes, you really can backdoor federated learning. In: NIPS (2020)
69. Warner, S.L.: Randomized response: a survey technique for eliminating evasive answer bias. JASA **60**(309), 63–69 (1965)
70. Yang, G., Wang, S., Wang, H.: Federated learning with personalized local differential privacy. In: IEEE ICCCS (2021)
71. Yang, Y., Dunlap, R., Rexroad, M., Cooper, B.F.: Performance of full text search in structured and unstructured peer-to-peer systems. In: INFOCOM (2006)
72. Zhang, C., Li, S., Xia, J., Wang, W., et al.: BatchCrypt: efficient homomorphic encryption for Cross-Silo federated learning. In: USENIX ATC (2020)
73. Zhang, Z., Wang, T., Li, N., He, S., et al.: CALM: consistent adaptive local marginal for marginal release under local differential privacy. In: ACM CCS (2018)

ANTM: Aligned Neural Topic Models for Exploring Evolving Topics

Hamed Rahimi[(✉)], Hubert Naacke, Camelia Constantin, and Bernd Amann

Sorbonne Université, CNRS, LIP6, 75005 Paris, France
{hamed.rahimi,hubert.naacke,camelia.constantin,
bernd.amann}@sorbonne-universite.fr

Abstract. In today's world, where the amount of textual information generated by humans and machines is rapidly growing, computational methods for organizing, summarizing, and tracking textual information and its evolution are becoming increasingly important. This paper introduces ANTM, an algorithmic family of dynamic topic models that combines novel techniques for discovering evolving topics in large corpora. ANTM preserves the temporal continuity of evolving topics by extracting temporal features from documents with advanced pre-trained large language models and by employing an overlapping sliding window algorithm for sequential aligned document clustering. This clustering method identifies different numbers of topics within each time frame and aligns semantically similar document clusters across time periods. It captures emerging and fading topics over time and allows for a more diverse and coherent representation of evolving topics. We evaluate ANTM against four other dynamic topic models on three datasets and conclude that it outperforms the state-of-the-art approaches in terms of interpretability and diversity. Furthermore, we demonstrate its effectiveness in handling large corpora, while improving the scalability and adaptability of dynamic topic models to different domains.

Keywords: Dynamic Topic Models · Aligned Neural Topic Models · Evolving Topics

1 Introduction

Topic modeling is a statistical technique used in natural language processing to discover abstract themes from a corpus of text documents [1,2]. Topic models are widely used in exploratory data analysis for organizing, understanding, and summarizing large amounts of text data [3]. We categorize topic models into Probabilistic Topic Models (PTM) [4–6] and Algorithmic Topic Models (ATM) [7–10] based on the methodologies used in their statistical modeling [11]. PTMs assume each document in a corpus is a mixture of topics, and each topic is a probability

---

 https://github.com/hamedR96/antm

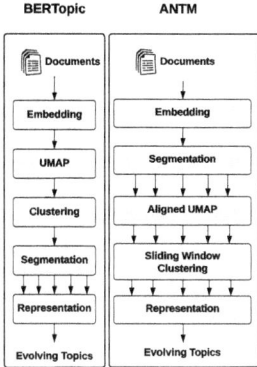

Fig. 1. BERTopic and ANTM

distribution over the words in the corpus. In contrast, ATMs exploit recent neural document and word representation models and take advantage of numerical optimization techniques to represent topics as weighted word vectors extracted from a set of semantically similar documents.

Dynamic topic models are the temporal variants of topic models that update their estimates of the underlying topics as new documents are added to the corpus [12,13]. They can be used to analyze topic evolution and to identify temporal patterns in the contents of document archives [14,15]. More specifically, they can be applied, for example, to discover the evolution of research topics and innovations in scientific archives [16] and to understand trends in public opinion on particular issues [17]. Dynamic topic models use different assumptions to statistically characterize the evolution of a document corpus in form of dynamic topics [18].

Like their static counterparts, dynamic topic models can be divided into Probabilistic Dynamic Topic Models (PDTM) and Algorithmic Dynamic Topic Models (ADTM). While PDTMs are widely used to analyse topic evolution [19, 20], they become computationally expensive when applied to large archives with extensive vocabularies [21]. In comparison to ADTMs, their lack of scalability is due to the need to sample from complex posterior distributions [22]. On the other hand, recent state-of-the-art ADTMs such as BERTopic [8], also suffer from certain limitations [23] due to the topic extraction process.

BERTopic. As depicted on the left of Fig. 1, BERTopic first computes low-dimensional semantic document embeddings with the Bert language representation model [24] (Embedding) and the UMAP dimension reduction algorithm [25]. It then applies a static clustering technique to extract topic document clusters (Clustering) and segments each document cluster along an overlapping temporal time window (Segmentation) before computing the term vector representation (Representation) with c-TF-IDF [26]. The transformation of static document cluster into dynamic topics through segmentation has an important drawback since there are instances where documents are categorised together under the same global topic, although in reality they belong to separate topics within a specific temporal frame [23]. This problem hampers the ability to capture the temporal variations of dynamic topics, including changes in the number of document clusters or the size of each cluster, as topics emerge and fade over time. We will show in our experiments that these limitations affect the quality in terms of topic diversity and topic coherence (Sect. 4.5).

To address the previous limitations of BERTopic, we introduce a new clustering-based dynamic topic model, called Aligned Neural Topic Model (ANTM). As depicted on the right of Fig. 1, ANTM applies a similar process as BERTopic but in a different order. In particular, it performs the temporal segmentation step to the document embeddings *before* the dimension reduction step (Aligned Umap). The document cluster segments are then reassembled along the time dimension (Sliding Window Clustering) to obtain dynamic topic document clusters. ANTM considers the changes over time in the document content, producing a range of high-quality topics for each period. By implementing an overlapping sliding window algorithm, it becomes possible to identify collections of related topics that cross multiple periods but remain distinct enough to demonstrate a form of evolution within a single domain [27]. Our experiments conducted on three datasets and four dynamic topic models demonstrate a considerable improvement in coherence and diversity scores compared to the SOTA PDTMs and ADTMs.

Contributions. We propose ANTM, a new family of dynamic topic models that effectively capture topic evolution using advanced algorithms. Second, we conduct a comprehensive analysis to evaluate the diversity and interpretability of the evolving topics generated by ANTM compared with state-of-the-art dynamic topic models, including recent ADTMs like D-ETM [28] and BERTopic [8], as well as traditional PDTMs such as TOT [14] and DTM [12]. Thirdly, we analyse various configurations of ANTM and compare their runtime and quality scores. Finally, we provide a qualitative analysis, limitations, and directions for future work.

Organization. The rest of the paper is organized as follows. Section 2 presents the state-of-the-art, Sect. 3 introduces ANTM, Sect. 4 presents the experimental details as well as datasets, baseline, evaluation metrics, and results. Finally, we conclude the paper in Sect. 5.

2 Related Work

Probabilistic Dynamic Topic Models. PDTMs [28,29] assign probabilities to different words and topics over time, allowing to infer the underlying themes and patterns in the data as they change. DTM (Dynamic Topic Model) [12] is one of the first PDTMs that incorporates temporal components for representing topic evolution within a topic model. DTM is a variant of the Latent Dirichlet Allocation (LDA) [4] model, which employs a Bayesian approach to characterize the evolution of the content within topic documents over time. DTM is extended by several approaches to deal with the temporal continuity of evolving topics in different ways [30,31]. The discrete-time Dynamic Topic Model (dDTM) [32] discretizes the data into time intervals, while the Continuous-Time Dynamic Topic Model (cDTM) [29] handles any data point in time, regardless of the time resolution. Topics Over Time (TOT) [14] is another PDTM that models time jointly with word co-occurrence patterns. TOT assumes that time is inherently continuous and can be characterized with a continuous distribution over timestamps [33]. Although PDTMs have been innovative tools for studying topic evolution, they have several limitations. Scalability is a first issue, as PDTMs become computationally expensive and time-consuming when dealing with very large dataset [34]. In addition, PDTMs cannot fully capture fine-grained content evolution because they expect that topics remain rather stable over time [35,36]. Several generative probabilistic topic models attempt to overcome these problems [37–40]. For example, Dynamic Embedded Topic Models (D-ETM) [28,41] combines DTM with word embeddings [42] to improve the performance of topic models by providing more informative word representation.Despite their benefits, D-ETMs remain computationally expensive and, as we will demonstrate in Sect. 4, their topic representations tend to suffer from lower diversity and coherence compared to those generated by ADTMs. As another example, to reduce the computational complexity and the number of variational parameters for handling large vocabularies, an amortized variational inference method based on DTM is presented in [43]. There are also efforts in the development of scalable PDTMs, such as [44], which extends the class of tractable priors from Wiener processes to the more general class of Gaussian processes. This approach allows the model to be applied to large text collections and to explore topics that evolve smoothly over a long time period.

Algorithmic Dynamic Topic Models. ADTMs [8,45,46] formalize the underlying probability distributions of words and documents into a fixed-length vector representation [47] using advanced data-mining techniques. This vector representation enables the application of various efficient clustering methods to build semantically coherent and diverse topic document clusters and topic representations. BERTopic [8] is a prevalent ADTM embedding context within the produced topics by using the Bert language representation model [24] for generating document embedding vector representations and c-TF-IDF [26] to produce topic representations from topic document contents. The vector representation allows to generate dense topic clusters of semantically coherent documents. As already discussed in Sect. 1 and shown in Fig. 1, BERTopic generates static topic clusters over the entire dataset and dynamic topic representations a-posteriori by applying c-TF-IDF on a sliding time window over each topic cluster.

The authors of [48] propose an assignment method to enhance topic descriptions. They consider the union of topic descriptions as a set of candidate words. Then, they add an optimisation step to select the 10 best words (i.e. maximizing a global relevance score) for each topic while ensuring that topic coverage and representation size is sufficient. This work has demonstrated improved topic

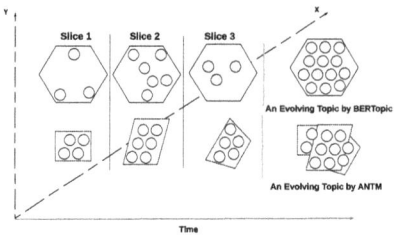

Fig. 2. Comparison of Evolving Topics in BERTopic and ANTM: ANTM's sliding window clustering method produces local document clusters for each time period which are aligned in a second step to produce global dynamic topics (sets of aligned local clusters). BERTopic performs global clustering and splits each topic document cluster according the different time periods, resulting in topic representations (obtained from the document contents) that fail to precisely portray evolving topics within a specific time frame.

quality for small texts (tweets) with limited vocabulary. It is orthogonal to ours and we might exploit it in future work.

Discussion. PDTMs lack of scalability due to the need to sample from complex posterior distributions [22] and state-of-the-art ADTMs such as BERTopic [8] suffer from certain limitations as discussed in [23] and in Sect. 1. The proposed method in this paper combines the benefits of using language models and efficient clustering algorithms to achieve scalability and the advantage of a dynamic topic cluster generation process of PDTMS for improving the quality of the generated topics.

3 ANTM

As illustrated in Fig. 3, the ANTM architecture consists of three layers. The first layer leverages advanced pre-trained language models (LLMs) to generate vector representations for each document, capturing its content (Document Vectors). The second layer, referred to as the SWS Splitter, subdivides the document vectors into temporally overlapping subsets (Temporal Document Vectors) and applies the AlignedUMAP algorithm [49] to produce low-dimensional document vector embeddings with global coherence for each subset. Subsequently, the hierarchical density-based clustering algorithm HDBSCAN [50] is applied to each overlapping subsets to generate topic clusters for each time period, which are then aligned to produce dynamic topic clusters (aligned subsets of clusters over all time periods). In the third layer, the Topic Representation Layer, we introduce a contextualized LLM-based Nearest Words approach to generate word representations for the aligned topic clusters. This layer can also incorporate class-based TF-IDF [26]. The following subsections provide a detailed definition and implementation of each layer.

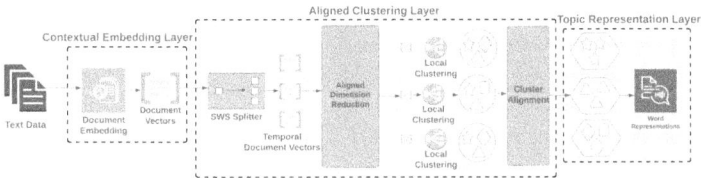

Fig. 3. The Architecture of ANTM: The first layer uses advanced pre-trained LLMs to provide a time-aware vector representation for each document. The second layer splits the document vector representations into a set of temporal time frames and performs an overlapping sliding window algorithm for temporal document clustering. Finally, the third layer is responsible for providing word representations for each set of aligned clusters over time.

3.1 Contextual Embedding Layer (CEL)

The Contextual Embedding Layer (**CEL**) is responsible for providing a vector representation for the document d in the corpus D. More formally, the embedding vector y for a document d is a mapping $\mathbf{CEL}: d \mapsto y \in \Re^z$, capturing contextual and semantic information from the corpus D. The document embedding represents words and documents in a low-dimensional feature vector space, where the embedding dimension z is expected to be much smaller than the size of the vocabulary (i.e., the number of unique words in D).

Implementation. CEL takes advantage of pre-trained LLMs (e.g., Data2Vec, GPT4) to compute a contextualized time-aware vector representation for each document. Pre-trained LLMs models capture the meaning and context of a document by using an attention mechanism to consider the context of words in a document. As shown in Eq. (1), given an input document d, the number u of tokens in d, and h_j the hidden state vector of the j-th token in d, the embedding $y = \mathrm{CEL}(d)$ of the whole document d is then obtained by taking the mean of all the hidden state vectors of the tokens in d.

$$\mathbf{CEL}: D \mapsto Y = [\mathbf{h}_z]_{z=1}^{u} = \frac{1}{u}\sum_{z=1}^{u} \mathbf{h}_z \tag{1}$$

We argue that LLM-based embeddings implicitly capture the temporal context of the document in addition to capturing its meaning. Indeed, the attention mechanism weights the importance of each word in the context of a document. This context implicitly depends on the publication period of the document since word contexts happen to vary over time.

3.2 Aligned Clustering Layer (ACL)

This layer is motivated by the belief that topic evolution occurs in cycles of steps as explained by [51]. Therefore, we discretize these steps by time frames with

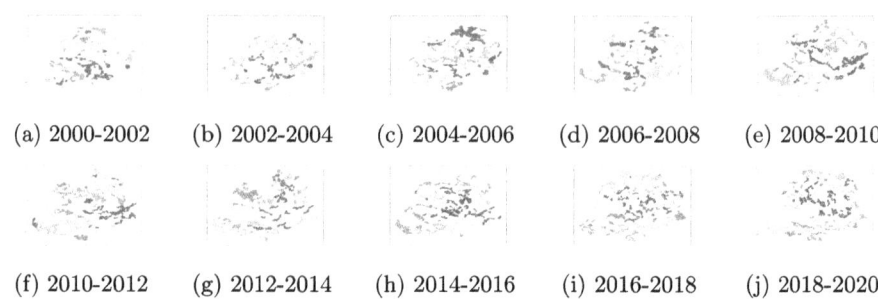

Fig. 4. Partitioned Clusters: The documents from DBLP dataset are embedded using BERT and sequentially clustered in each time frame. The clusters of all periods are aligned to create evolving topics as described in Fig. 5.

intersections similar to [12,15], but we use time-aware algorithms to account for the continuity of time and to model evolving topics that emerge and fade over time. After performing transformer-based document embedding on the corpus D, we can obtain a dynamic vector representation of D by applying a Sliding Window Segmentation (**SWS**) process that divides D into a series of n overlapping time frames $\{W^1, \ldots, W^n\}$. We then denote the set of documents published in the time frame W^t by $D^t \subseteq D$ and the set of embeddings of D^t by $Y^t = \{\mathbf{CEL}(d) | d \in D^t\}$. This layer aims to discover evolving document clusters by sequentially grouping similar documents. This procedure is called Partitioned Clustering (**PC**).

Definition 1. *Let D^t be a set of documents in the t-th time frame of the document corpus D and let Y^t denote the set of embeddings of D^t. Partitioned Clustering $\mathbf{PC}: D^t \mapsto \{D_i^t\}$ clusters the documents D^t by their embeddings Y_t and returns a set of local document clusters $D_i^t \subseteq D$ for $i = 1$ to k_t where all document embeddings in Y_i^t are similar. Y_i^t is the set of embeddings of the documents contained in cluster D_i^t.*

Figure 4 shows two-dimensional document embedding vectors clustered by their cosine distance for different time periods. The sequence of figures reveals evolution patterns and trends of document clusters (topics) that may not be apparent when documents are grouped based on their content.

To obtain a representation of the temporal evolution of these clusters, we apply a final step that aligns the document clusters along all periods using a cluster linkage measure. We can use different cluster linkage measures (single, average, centroid, complete) to estimate the similarity of the vector embedding clusters Y_i^t for local document clusters D_i^t. The result of each alignment is a set of *evolving document clusters* D_k (EC) that contain semantically similar documents from different time periods:

Definition 2. *Let $\{D_i^t\}$ be the set of local document clusters obtained by $\mathbf{PC}$. Cluster Alignment $\mathbf{CA}: D \mapsto \{D_k\}$ generates m subsets of documents $D_k \subseteq D$*

Fig. 5. Evolving Clusters: All documents within an evolving cluster share the same color and are the results of a local clustering step for each time period (x-y plane) followed by a cluster alignment step along time (z-axis). We can see that aligned clusters which have the same color are rather vertically aligned through consecutive time frames ("slices"). The representation of these Evolving Clusters can help explore the evolution of topics over time as described in Fig. 10, and Fig. 9.

for $k = 1$ to m and where D_k is the union of a set of local document clusters (topics) D_i^t with similar embedding clusters Y_i^t. Figure 5 shows the evolving clusters after the alignment.

Implementation. The Aligned Clustering Layer (ACL), similar to the other two layers of ANTM, can be implemented using different clustering and alignment methods combined with scalable algorithms. As shown in Fig. 3, we implemented the ACL layer with density-based clustering techniques [52] to cluster data points (embedding vectors) based on their distribution in the feature space. These algorithms are particularly useful for identifying clusters of varying densities and arbitrary shapes. The number of obtained clusters are defined by the parameters used to estimate the density. This feature provides an advantage over non-parametric PDTMs that require a fixed number of topics as input. The Aligned Clustering Layer is developed in the following three steps.

Aligned Dimension Reduction. To overcome the curse-of-dimensionality problem of similarity based clustering methods on high dimensional data, we first reduce the size of document embedding vectors. The proposed solution uses AlignedUMAP [53] to align sequences of UMAP [25] embeddings based on the documents shared between two consecutive time periods. These overlapping documents act as landmarks for aligning the different vector spaces defined for each period and allow us to obtain globally coherent reduced-dimension vector representations.

Sequential Local Clustering. After the aligned dimension reduction, a density-based clustering algorithm is performed on each period data frame to separate all documents of a given period into a set of local document clusters. We use HBDSCAN [50], a hierarchical version of DBSCAN [54] which does not require a predefined number of clusters but generates and selects clusters based on a notion of stability (the hierarchical clustering process slices the cluster hierarchy so that the number of clusters is as close as possible to that of the next level in the hierarchy, while constraining the cluster size to discard clusterings with too large and too few clusters). As shown in Fig. 4, each such local cluster describes a concept within a time frame.

Cluster Alignment. The following step involves aligning consecutive local clusters that were achieved in the previous step. As the document embedding vectors of all time frames are in the same space, we can aggregate each cluster by considering their centroids. We use the centroid linkage method [55] to align document clusters from different periods by clustering the cluster centroids using HDBSCAN. An example of the output for this step is shown in Fig. 5. An other alternative for HDBSCAN is to use the K-Nearest Neighbors (KNN) clustering method based on cosine similarity.

KNN clustering allows us to introduce a similarity threshold parameter to determine the strength of the alignment links. A high threshold results in fewer alignments, highlighting topics that barely change over time. On the other hand, a low threshold produces more alignments, including topics with significant changes over time and lower similarity values. This enhances the versatility of the topic alignment step and helps to improve the representation quality of evolving topics. It also helps to avoid aligning topics of the same time periods and to focus only on aligning the topics of each time period with their adjacent time periods. Figure 6 illustrates an evolving topic about the Ebola outbreak using HDBSCAN. We can see that the topic representation (defined in Sect. 3.3) changes slightly over time, and most of the topic words are stable and repeated in the next time period. At time period 1, every word is considered as new (green color). For the following periods, a word contained in the previous period and the next one is considered as stable (grey color), a word not in the next period is considered as disappearing (red color). However, if we align topics using KNN which is more flexible that HDBSCAN, we can obtain a richer evolving topic containing a larger set of words about various diseases, see Fig. 7 where the similarity threshold has been set to 0.6 which is rather low. If we increase the similarity threshold from 0.6 to 0.8, the alignment in Fig. 7 splits into multiple smaller alignments that showcase the evolution of specific diseases and the Ebola outbreak is one of them (Fig. 8). This increases the versatility of the topic alignment process, which can be configured based on application and user preferences.

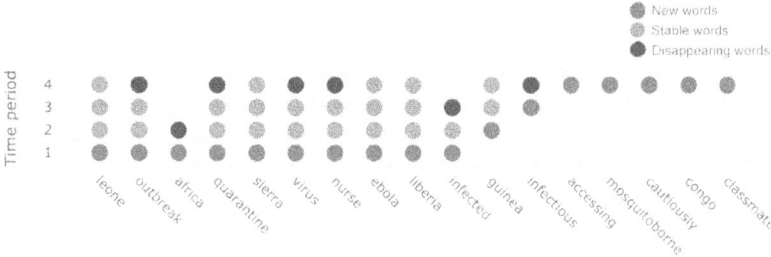

Fig. 6. An evolving topic regarding Ebola Outbreak based on HDBSCAN topic alignment. (Color figure online)

Fig. 7. Evolving topic about diseases and outbreaks - Alignment using KNN with threshold = 0.6

3.3 Representation Layer

The Representation Layer (RL) is responsible for generating word representations for each local topic document cluster in each time frame. The goal of this layer is to identify the most relevant terms or term phrases to summarize the main ideas or themes of each document cluster.

Definition 3. *The topic Representation Layer (**RL**) computes a list of m terms $\{t_{ij}^r\}_{r=1}^m$ that describe the semantic contents of each document cluster D_{ij}.*

Implementation. There exist various ways to represent a set of documents by a set of terms [48]. One way is to use a method called class-based Term Frequency-Inverse Document Frequency (c-TF-IDF) [8,26], which is a variation of TF-IDF that takes into account the class labels of documents. c-TF-IDF weights the terms in a cluster not only by their frequency but also by their relevance to a particular cluster. c-TF-IDF ensures that each word of a topic representation occurs in at least one document of that topic. Examples of outputs for this step are shown in Figs. 9 and 10. We propose to use an LLM-based Nearest Words approach similar to [10] that computes joint word and document embeddings over a document cluster and finds the m closest words from the centroid of the document embeddings. LLM-based Nearest Words generates a vector representation for each document by averaging its word vectors. Subsequently, it derives a contextualized vector representation for each word through max-pooling across all available vectors for that word across all documents. Max pooling captures the diversity of various contexts of a given word across different dimension.

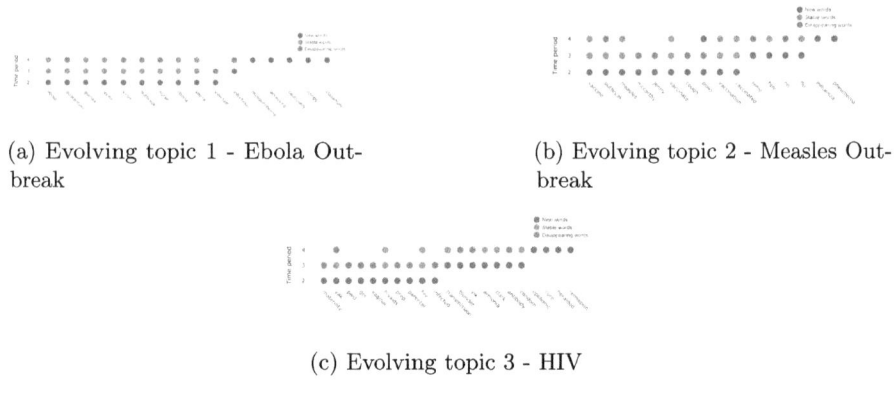

(a) Evolving topic 1 - Ebola Outbreak

(b) Evolving topic 2 - Measles Outbreak

(c) Evolving topic 3 - HIV

Fig. 8. Topic alignment using KNN with 0.8 unit threshold

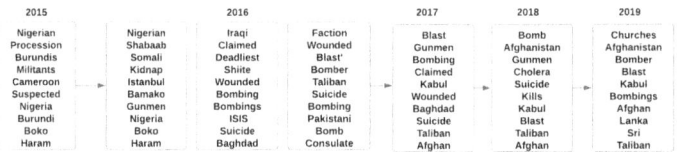

Fig. 9. Evolution of New York Times News on Foreign Terrorist Activities. The interesting transition of news from Boko, Harram, Nigeria in 2015 to ISIS, Iraq, Baghdad in 2016 and later Kabul, Taliban, Afghanistan in 2017 to 2019.

Fig. 10. Evolution of Computer Science Research on Medical Science based on DBLP documents. The words CNN, Lung, and Diagnosis appear in the period 2018 to 2020, which coincides with the pandemic and advances in computer vision research.

As shown in Fig. 11, the proposed topic representation method achieves higher median coherence compared to c-TF-IDF and the non-contextualized Nearest Words representations, which use generic word embeddings without considering the context of the words in each document, as proposed in [10].

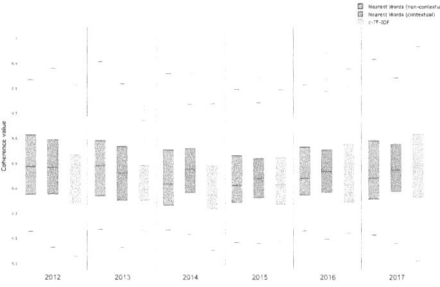

Fig. 11. Coherence value distributions of non-contextual Nearest Words, contextual Nearest Words, and c-TF-IDF

Table 1. Datasets

Dataset	Documents	Tokens	Vocabulary	Date Range
DBLP	200K	18.5M	144K	2000-2020
arXiv	70K	1.17M	12.6K	2000-2022
NYT News	210K	5M	61K	2012-2022

4 Experiments

We compare ANTM with four other dynamic topic models on three datasets. Our experiments aim to demonstrate ANTM's performance in terms of topic coherence and diversity relative to state-of-the-art competitors and to illustrate its capability for exploratory topic evolution analysis. Besides, we provide a comparative study that examines ANTM under different settings for each layer in different scenarios.

4.1 Datasets

We use three datasets in the experiments. The first dataset is the DBLP [56] archive of 200K scientific articles (title and abstract) published between 2000 and 2020. DBLP is an open bibliographic database, search engine, and knowledge graph that archives computer science publications. The second dataset is a 70K sample of documents (title and abstract) extracted from arXiv [57] and published between 2000 and 2022. The last dataset is a collection of 210K articles [58] from the New York Times (NYT) published between 2012 and 2022. The NYT dataset includes historical and current articles on a variety of topics and includes text, images, and metadata such as the article's headline, author, and publication date. The statistics of the datasets are summarized in Table 1.

4.2 Baseline Models

The performance of ANTM is compared to four other dynamic topic models. The first model is DTM [12], which has been widely used for various applications such as public opinion tracking [59]. The second model is D-ETM [28], which extends DTM with word embeddings to improve topic quality and performance. The third topic model is TOT [14] that is a variation of LDA and captures how the co-occurrence patterns of words changes over time. The last model is BERTopic [8], a clustering-based topic model that uses a static process for clustering document embeddings while dynamically representing topics.

4.3 Evaluation Metrics

We evaluate the performance of each topic model in terms of human interpretability using Topic Coherence (TC) [60–62] and its diversity by Topic Diversity (TD) [28,63].

Topic Coherence (TC). This metric indicates the interpretability of a topic containing m words and by considering that a topic is highly coherent if it is represented by words that tend to occur in the same documents of a collection. TC, as defined in [61] with co-occurrence value C_V, is a variation of Normalized Pointwise Mutual Information (NPMI) [64] that averages the co-occurrence values of word pairs (t_i^r, t_i^s) in all topicss i over N topics.

Topic Diversity (TD). This metric estimates the diversity of the topic representations within a given set of topics and the facility to distinguish different topics from each other by their representation. For this metric, we use the Proportion of Unique Words (PUW) method [41] defined as the vocabulary size divided by the total number of words within a set of topics. Topics within a low-diversity topic set share many words, whereas a highly-diverse topic set contains topics that have few words in common.

4.4 Experimental Setup

ANTM. We used two embedding models from SBERT [65] in the contextual embedding layer (**CEL**) of ANTM. The two configurations with these models are called ANTM-large (using *all-mpnet-base-v2*) and ANTM-mini (using *all-MiniLM-L6-v2*). We also used the Data2Vec pre-trained model (*facebook/data2vec-text-base*) [66] in our study. We then split the document embedding vectors into a series of time frames to explore the change of document contents. The setting of the segmentation step is summarized in Table 2. These segmentation values, as suggested in [67], provide a comprehensive view of the data and ensure that changes over time are captured in the analysis.

Afterwards, we chose the hyperparameters for dimension reduction (using AlignedUMAP) and sequential document clustering (using **PC**). As suggested in [8,10,49], the cosine similarity metric, with 5-dimensional output, was chosen

Table 2. Segmentation setting

Dataset & Time Frame Sizes	Length	Overlap	#Frames
DBLP Documents	3 Years	1 year	10
arXiv Documents	4 Years	1 year	10
NYT News	4 Years	1 year	10

for the dimensionality reduction setting of AlignedUMAP. We then performed HDBSCAN on each time frame with Euclidean distance and a minimum size of 10 documents per cluster to create a set of semantically similar document clusters for each period (Fig. 4). Since all document embeddings are in the same vector space, we could then align the generated clusters by again using HDBSCAN on the centroid of all document clusters with Euclidean distance and a minimum number of 2 clusters to obtain evolving clusters for each dataset (Fig. 5). Finally, we represented the documents of each cluster with a set of $m = 10$ words using the c-TF-IDF (Figs. 9 and 10). The common parameters and methods are chosen similarly to BERTopic for fair comparison.

Baselines. DTM, D-ETM and TOT were configured with 20, 50 and 100 topics respectively and applied to the titles and abstracts of the DBLP, arXiv, and NYT News datasets. The topic numbers were determined based on the numbers of topics generated by ANTM in an unsupervised manner. Additionally, we ran BERTopic using the same sentence transformer models as the proposed model for a fair comparison. The default configurations were used for the rest of the hyperparameters of BERTopic.

4.5 Results

The performance of dynamic topic models can be analyzed within two perspectives. First, we observe the quality of topics in terms of coherence and diversity within each time frame (period-wise analysis). This aims to assess the ability of dynamic topic models to describe temporal topics in a diverse and coherent manner. The second perspective analyzes the quality of word representations within each evolving topic (topic-wise analysis) and compares the ability of dynamic topic models to represent the evolution of each dynamic topic over time.

As shown in Table 3, ANTM achieves the highest Topic Quality (TC×TD) scores among the different baseline variants in all three datasets. However ANTM is 2.8 to 4.5 slower than BERTopic depending on the used embedding model and due to the increased number of topic clustering steps. Yet, ANTM's runtime for producing 62 topics is respectively 8, 30 and 43 times faster than its competitors when generating 50 topics.

Period-Wise Quality Comparison. The goal of the period-wise analysis is to determine whether aligned document clustering, applied separately for each time frame, results in better topic quality per period. As shown in Fig. 12, ANTM

Table 3. Performance comparison of ANTM and baselines

Topic Model	Embedding Model	ArXiv					DBLP					NY Times				
		#T	TC	TD	TQ	t(s)	#T	TC	TD	TQ	t(s)	#T	TC	TD	TQ	t(s)
DTM	-	20	0.56	0.71	0.40	16194	20	0.58	0.68	0.40	15485	20	0.57	0.95	0.54	3071
	-	50	0.61	0.72	0.44	23897	50	0.61	0.74	0.45	33541	50	0.46	0.98	0.45	6535
	-	100	0.61	0.79	0.48	46195	100	0.64	0.77	0.49	61140	100	0.38	0.99	0.37	13255
TOT	-	20	0.33	0.15	0.05	14071	20	0.33	0.07	0.03	20385	20	0.42	0.13	0.06	2680
	-	50	0.32	0.07	0.02	15833	50	0.33	0.04	0.02	45295	50	0.41	0.15	0.06	6490
	-	100	0.32	0.07	0.02	29205	100	0.33	0.05	0.02	91725	100	0.38	0.14	0.06	12146
D-ETM	Word2Vec	20	0.45	0.88	0.40	2925	20	0.44	0.85	0.37	12352	20	0.48	0.98	0.47	8257
	Word2Vec	50	0.47	0.69	0.32	4625	50	0.49	0.71	0.34	19332	50	0.43	0.79	0.34	12907
	Word2Vec	100	0.49	0.48	0.23	6840	100	0.45	0.61	0.28	36471	100	0.38	0.52	0.20	20133
BERTopic	all-MiniLM-L6-v2	681	0.68	0.88	0.60	**192**	1075	0.70	0.78	0.54	**342**	1539	0.60	0.91	0.54	**340**
	all-mpnet-base-v2	812	0.70	0.89	0.63	**90**	1472	0.71	0.81	0.58	**311**	1408	0.63	0.92	0.57	**143**
ANTM	all-MiniLM-L6-v2	62	0.72	0.94	0.67	544	130	0.77	0.91	0.70	1398	118	0.62	0.93	0.57	1549
	all-mpnet-base-v2	172	0.59	0.92	0.54	405	261	0.70	0.93	0.65	1869	104	0.63	0.92	0.57	1660

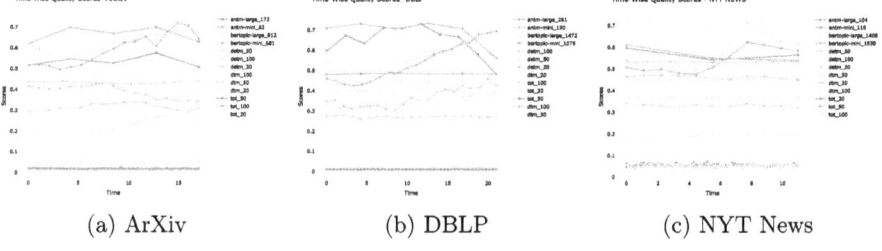

(a) ArXiv (b) DBLP (c) NYT News

Fig. 12. Period-wise Quality Comparison

overall has higher quality scores compared to D-ETM, DTM, and TOT in all three datasets. In contrast to BERTopic, which exhibits ascending values across consecutive time frames, our proposed method maintains a stable consistency. As illustrated in Fig. 2, the variability observed in BERTopic's performance stems from its global clustering approach with uniform parameters, resulting in uneven document distributions within each cluster. It is noteworthy, though, that the quality of ANTM experienced a decrease in the last time frame. We argue that this decrease can be attributed to the data scarcity for the last time frame within the datasets. Moreover, we observe that DTM generates topics with higher consistence compared to D-ETM, even without the use of embedding models. Lastly, it is worth noting that TOT produces the lowest-quality evolving topics across all the datasets.

Topic-Wise Quality Comparison. The objective of topic-wise analysis is to assess whether an evolving topic, which contains topics that are semantically close to each other, can nevertheless effectively represent the evolution of the represented document clusters. As shown in Fig. 13, topic quality (TQ) is calculated for each evolving topic and their distributions is plotted across each model on all three datasets. These figures lead to various observations.

First, as depicted in Fig. 13(b) and (c), ANTM consistently generates the highest quality evolving topics in both the DBLP and NYT News datasets. However, in the case of ArXiv, BERTopic outperforms the ANTM-large model in terms of topic-wise quality, although it still falls short of the quality achieved

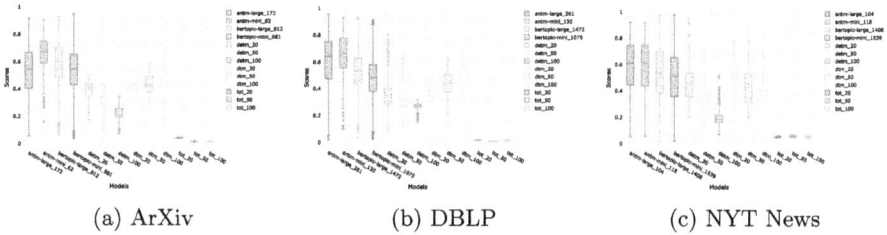

(a) ArXiv (b) DBLP (c) NYT News

Fig. 13. Topic-wise Quality Comparison

by the ANTM-mini model. The second observation highlights that DTM consistently generates higher quality topics compared to D-ETM, even without the use of embedding models. Lastly, it is worth noting that TOT consistently produces the lowest quality evolving topics across all the datasets.

Table 4. Variation Comparison of ANTM

Contextualized Embedding Layer	Aligned Clustering Layer				Representation Layer					
	SWS		Sliding Window Clustering		c-TF-IDF					
Embedding Model	Window length	Overlap length	AlignedUMAP #Neighbor	HDBSCAN Min Cluster	#Words	#T	TC	TD	TQ	t(s)
all-MiniLM-L6-v2	5	2	25	25	5	60	0.74	0.93	0.70	747
					10	60	0.61	0.89	0.55	752
	4	1	20	20	5	66	0.71	0.93	0.66	**474**
					10	66	0.56	0.90	0.51	**478**
all-mpnet-base-v2	5	2	25	25	5	75	0.74	0.92	0.68	740
					10	75	0.59	0.88	0.52	806
	4	1	20	20	5	80	0.68	0.92	0.63	505
					10	80	0.54	0.89	0.48	547
data2vec	5	2	25	25	5	10	0.40	0.65	0.26	801
					10	10	0.30	0.60	0.18	823
	4	1	20	20	5	8	0.39	0.77	0.30	503
					10	8	0.30	0.73	0.22	531

Variation Analysis. As shown in Table 4, we compare different configurations of ANTM, similar to an ablation study, by systematically varying the values of one or more algorithm parameters while keeping other aspects constant. By observing how changes in parameter values affect the performance of the model, we can identify the most influential parameters and their optimal settings. In this comparative analysis, we have selected three distinct language models for the contextualized embedding layer, allowing us to investigate the impact of pretrained models. Additionally, we have explored two distinct configurations for the aligned clustering layer, facilitating a comparison between scenarios involving long time frames with substantial overlap and those with shorter time frames and reduced overlap. Furthermore, we have examined two distinct settings for the representation layer to assess the model's representational capacity concerning the number of words. The details of this comparative analysis are provided in Table 4. In our findings, it becomes evident that ANTM's performance significantly improves when utilizing BERT as the embedding layer. Moreover,

Table 5. Qualitative Comparison

	DTM	TOT	D-ETM	BERTopic	ANTM
Temporal Labeling	✓	✗	✓	✓	✓
Non-parametric	✗	✗	✗	✓	✓
Semantization	✗	✗	Word2Vec	LLMs	LLMs
Coherence	Normal	Very Low	Low	High	Very High
Diversity	Low	Low	Normal	High	Very High
Runtime	Very High	High	Normal	Very Low	Low
Complexity	Low	Low	Normal	High	High
Scalability	Very Low	Very Low	Low	Normal	High
Adaptability	Very Low	Very Low	Low	High	High

employing larger time frames with higher overlap yields superior results, and intriguingly, a representation layer comprising five words outperforms one with ten words.

Qualitative Comparison. The qualitative comparison in topic modeling is crucial for making sense of the results, refining models, validating them against domain knowledge, and ultimately extracting meaningful insights from textual data. While quantitative metrics provide valuable guidance, qualitative assessment is essential for ensuring the relevance and human interpretability of the topics generated by these models. As shown in Table 5, a qualitative comparison between ANTM and the baselines is provided. This comparison includes *Temporal Labeling* (the ability to represent each evolving topic corresponding to a time period), *Semantization* (the ability to understand the semantics behind the text representation), *Topic Coherence* and *Diversity* of labels as discussed in Table 3, *Runtime* (the computational efficiency of each model, considering the time required to process large datasets and generate topic models), algorithmic *Complexity* (including the number of hyperparameters and algorithmic complexity), *Scalability*, which assesses the ability to handle large corpora, and *Adaptability* with different algorithms concerning the data and the domain. These aspects are informed by our methodological understanding of different topic modeling approaches, which has allowed us to identify the key dimensions along which these models can be compared and contrasted. By considering these qualitative aspects, researchers and practitioners can gain a deeper understanding of the strengths and limitations of different topic modeling approaches, ultimately selecting the most suitable model for their specific use case.it

Limitation and Future Work. Initial experiments have shown that the hyperparameters need to be chosen carefully to avoid the formation of large topics that group together unrelated topics during certain time periods, as shown in

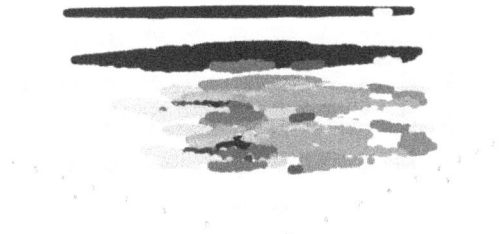

Fig. 14. The formation of non-informative clusters

Fig. 14. In future work, we aim to address these issues and plan to define a metric for assessing the quality of topic alignments and their ability to detect topic emergence or fading.

5 Conclusion

Existing dynamic topic models ignore certain temporal variations of evolving topics by configuring a global structure for dynamic topics, such as the same number of document clusters in each period. Furthermore, many of these models necessitate computationally expensive operations when dealing with large-scale corpora, which can be a significant bottleneck. These limitations directly affect the observation of topic evolution and reduce the coherence and diversity of evolving topics, which are sequentially represented. In this paper, we proposed a family of dynamic neural topic models called Aligned Neural Topic Models (ANTM), which combines novel data mining algorithms to provide a modular framework for discovering evolving topics. Based on a series of experiments on three distinct datasets we can conclude that ANTM outperforms the state-of-the-art dynamic topic models in terms of topic coherence and diversity.

Acknowledgment. We gratefully acknowledge the Sorbonne Center for Artificial Intelligence (SCAI) for partially funding this research through a doctoral fellowship grant. We'd like to thank our students, Lucie Chen, Mohamed Allaa Eddine Boutaleb, and Mouloud Samy Nehlil for their assistance.

References

1. Blei, D.M.: Probabilistic topic models. Commun. ACM **55**(4), 77–84 (2012)
2. Alghamdi, R., Alfalqi, K.: A survey of topic modeling in text mining. Int. J. Adv. Comput. Sci. Appl. (IJACSA) **6**(1) (2015)
3. Churchill, R., Singh, L.: The evolution of topic modeling. ACM Comput. Surv. **54**(10s), 1–35 (2022)
4. Blei, D.M., Ng, A.Y., Jordan, M.I.: Latent dirichlet allocation. J. Mach. Learn. Res. **3**(Jan), 993–1022 (2003)

5. Teh, Y., Jordan, M., Beal, M., Blei, D.: Sharing clusters among related groups: hierarchical dirichlet processes. In: Advances in Neural Information Processing Systems, vol. 17 (2004)
6. Steyvers, M., Griffiths, T.: Probabilistic topic models. In: Handbook of Latent Semantic Analysis, pp. 439–460. Psychology Press (2007)
7. Thompson, L., Mimno, D.: Topic modeling with contextualized word representation clusters, arXiv preprint arXiv:2010.12626 (2020)
8. Grootendorst, M.: Bertopic: neural topic modeling with a class-based TF-IDF procedure, arXiv preprint arXiv:2203.05794 (2022)
9. Bahrainian, S.A., Jaggi, M., Eickhoff, C.: Self-supervised neural topic modeling. In: Findings of the Association for Computational Linguistics: EMNLP 2021, Punta Cana, Dominican Republic, pp. 3341–3350. Association for Computational Linguistics (2021)
10. Angelov, D.: Top2vec: distributed representations of topics, arXiv preprint arXiv:2008.09470 (2020)
11. Breiman, L.: Statistical modeling: the two cultures (with comments and a rejoinder by the author). Stat. Sci. **16**(3), 199–231 (2001)
12. Blei, D.M., Lafferty, J.D.: Dynamic topic models. In: Proceedings of the 23rd International Conference on Machine Learning, pp. 113–120 (2006)
13. Abdelrazek, A., Eid, Y., Gawish, E., Medhat, W., Hassan, A.: Topic modeling algorithms and applications: a survey. Inf. Syst. **112**, 102131 (2023)
14. Wang, X., McCallum, A.: Topics over time: a non-markov continuous-time model of topical trends. In: Proceedings of the 12th ACM SIGKDD International Conference on Knowledge Discovery and Data Mining, pp. 424–433 (2006)
15. Hu, J., Sun, X., Lo, D., Li, B.: Modeling the evolution of development topics using dynamic topic models. In: 2015 IEEE 22nd International Conference on Software Analysis, Evolution, and Reengineering (SANER), pp. 3–12. IEEE (2015)
16. Li, K., Naacke, H., Amann, B.: An analytic graph data model and query language for exploring the evolution of science. Big Data Res. **26**, 100247 (2021)
17. Sha, H., Hasan, M.A., Mohler, G., Brantingham, P.J.: Dynamic topic modeling of the covid-19 twitter narrative among us governors and cabinet executives, arXiv preprint arXiv:2004.11692 (2020)
18. Zhou, H., Yu, H., Hu, R.: Topic evolution based on the probabilistic topic model: a review. Front. Comput. Sci. **11**, 786–802 (2017)
19. Greene, D., Cross, J.P.: Exploring the political agenda of the European parliament using a dynamic topic modeling approach. Polit. Anal. **25**(1), 77–94 (2017)
20. Yao, F., Wang, Y.: Tracking urban geo-topics based on dynamic topic model. Comput. Environ. Urban Syst. **79**, 101419 (2020)
21. Bhadury, A., Chen, J., Zhu, J., Liu, S.: Scaling up dynamic topic models. In: Proceedings of the 25th International Conference on World Wide Web, pp. 381–390 (2016)
22. Zhang, D.C., Lauw, H.: Dynamic topic models for temporal document networks. In: Chaudhuri, K., Jegelka, S., Song, L., Szepesvari, C., Niu, G., Sabato, S. (eds.) Proceedings of the 39th International Conference on Machine Learning. Proceedings of Machine Learning Research, vol. 162, pp. 26281–26292. PMLR (2022)
23. Eskonen, J.: Dynamic topic modeling and clustering: dynamic topic modeling and clustering of occupational health and safety publications. Master's thesis, Tampere University (2022)
24. Devlin, J., Chang, M.-W., Lee, K., Toutanova, K.: Bert: pre-training of deep bidirectional transformers for language understanding, arXiv preprint arXiv:1810.04805 (2018)

25. McInnes, L., Healy, J., Melville, J.: UMAP: uniform manifold approximation and projection for dimension reduction, arXiv preprint arXiv:1802.03426 (2018)
26. MaartenGr: cTFIDF - class-based TF-IDF implementation in Python (2022). https://github.com/MaartenGr/cTFIDF. Accessed 28 Aug 2023
27. Rahimi, H., Naacke, H., Constantin, C., Amann, B.: ATEM: a topic evolution model for the detection of emerging topics in scientific archives, arXiv preprint arXiv:2306.02221 (2023)
28. Dieng, A.B., Ruiz, F.J., Blei, D.M.: The dynamic embedded topic model, arXiv preprint arXiv:1907.05545 (2019)
29. Wang, C., Blei, D., Heckerman, D.: Continuous time dynamic topic models, arXiv preprint arXiv:1206.3298 (2012)
30. Iwata, T., Yamada, T., Sakurai, Y., Ueda, N.: Online multiscale dynamic topic models. In: Proceedings of the 16th ACM SIGKDD International Conference on Knowledge Discovery and Data Mining, pp. 663–672 (2010)
31. Ren, L., Dunson, D.B., Carin, L.: The dynamic hierarchical dirichlet process. In: Proceedings of the 25th International Conference on Machine Learning, pp. 824–831 (2008)
32. Bahrainian, S.A., Mele, I., Crestani, F.: Modeling discrete dynamic topics. In: Proceedings of the Symposium on Applied Computing, pp. 858–865 (2017)
33. Vayansky, I., Kumar, S.A.: A review of topic modeling methods. Inf. Syst. **94**, 101582 (2020)
34. Gillenwater, J., Kulesza, A., Taskar, B.: Discovering diverse and salient threads in document collections. In: Proceedings of the 2012 Joint Conference on Empirical Methods in Natural Language Processing and Computational Natural Language Learning, pp. 710–720 (2012)
35. Liu, Y., Wang, J., Qian, Y., Jiang, Y., Sun, J., Chai, J.: Dynamic topic model for tracking topic evolution and measuring popularity of scientific literature. In: 2021 IEEE Sixth International Conference on Data Science in Cyberspace (DSC), pp. 315–320. IEEE (2021)
36. Churchill, R.: Percolation-based topic modeling for tweets. In: KDD Conference (WISDOM 2020), San Diego, CA, USA (2020)
37. Wei, X., Sun, J., Wang, X.: Dynamic mixture models for multiple time-series. In: IJCAI, vol. 7, pp. 2909–2914 (2007)
38. Iwata, T., Yamada, T., Sakurai, Y., Ueda, N.: Online multiscale dynamic topic models. In: Proceedings of the 16th ACM SIGKDD International Conference on Knowledge Discovery and Data Mining, KDD 2010, pp. 663-672. Association for Computing Machinery, New York (2010)
39. Bhadury, A., Chen, J., Zhu, J., Liu, S.: Scaling up dynamic topic models. In: Proceedings of the 25th International Conference on World Wide Web, WWW 2016, Republic and Canton of Geneva, CHE, pp. 381–390. International World Wide Web Conferences Steering Committee (2016)
40. Zosa, E., Granroth-Wilding, M.: Multilingual dynamic topic model. In: RANLP 2019-Natural Language Processing a Deep Learning World (2019)
41. Dieng, A.B., Ruiz, F.J., Blei, D.M.: Topic modeling in embedding spaces. Trans. Assoc. Comput. Linguist. **8**, 439–453 (2020)
42. Mikolov, T., Sutskever, I., Chen, K., Corrado, G.S., Dean, J.: Distributed representations of words and phrases and their compositionality. In: Advances in Neural Information Processing Systems, vol. 26 (2013)

43. Tomasi, F., Lalmas, M., Dai, Z.: Efficient inference for dynamic topic modeling with large vocabularies. In: Cussens, J., Zhang, K. (eds.) Proceedings of the Thirty-Eighth Conference on Uncertainty in Artificial Intelligence. Proceedings of Machine Learning Research, vol. 180, pp. 1950–1959. PMLR (2022)
44. Jähnichen, P., Wenzel, F., Kloft, M., Mandt, S.: Scalable generalized dynamic topic models. In: Storkey, A., Perez-Cruz, F. (eds.) Proceedings of the Twenty-First International Conference on Artificial Intelligence and Statistics. Proceedings of Machine Learning Research, vol. 84, pp. 1427–1435. PMLR (2018)
45. Gao, Q., Huang, X., Dong, K., Liang, Z., Wu, J.: Semantic-enhanced topic evolution analysis: a combination of the dynamic topic model and word2vec. Scientometrics **127**(3), 1543–1563 (2022)
46. Eklund, A., Forsman, M., Drewes, F.: Dynamic topic modeling by clustering embeddings from pretrained language models: a research proposal. In: Proceedings of the 2nd Conference of the Asia-Pacific Chapter of the Association for Computational Linguistics and the 12th International Joint Conference on Natural Language Processing: Student Research Workshop, pp. 84–91 (2022)
47. Le, Q., Mikolov, T.: Distributed representations of sentences and documents. In: International Conference on Machine Learning, pp. 1188–1196. PMLR (2014)
48. Gracianne, O., Halftermeyer, A., Dao, T.: Presenting an event through the description of related tweets clusters. In: Reformat, M.Z., Zhang, D., Bourbakis, N.G. (eds.) 34th IEEE International Conference on Tools with Artificial Intelligence, ICTAI 2022, Macao, China, 31 October–2 November 2022, pp. 1283–1290. IEEE (2022)
49. How to use alignedumap
50. Campello, R.J.G.B., Moulavi, D., Sander, J.: Density-based clustering based on hierarchical density estimates. In: Pei, J., Tseng, V.S., Cao, L., Motoda, H., Xu, G. (eds.) PAKDD 2013. LNCS (LNAI), vol. 7819, pp. 160–172. Springer, Heidelberg (2013). https://doi.org/10.1007/978-3-642-37456-2_14
51. Kuhn, T.S.: The Structure of Scientific Revolutions. University of Chicago Press, Chicago (2012)
52. Ghosal, A., Nandy, A., Das, A.K., Goswami, S., Panday, M.: A short review on different clustering techniques and their applications. Emerg. Technol. Model. Graph. Proc. IEM Graph **2018**, 69–83 (2020)
53. Islam, M.T., Fleischer, J.W.: Manifold-aligned neighbor embedding, arXiv preprint arXiv:2205.11257 (2022)
54. Ester, M., Kriegel, H.-P., Sander, J., Xu, X., et al.: A density-based algorithm for discovering clusters in large spatial databases with noise. In: KDD, vol. 96, pp. 226–231 (1996)
55. Jarman, A.M.: Hierarchical cluster analysis: comparison of single linkage, complete linkage, average linkage and centroid linkage method. Georgia Southern University (2020)
56. Ley, M.: The DBLP computer science bibliography: evolution, research issues, perspectives. In: Laender, A.H.F., Oliveira, A.L. (eds.) SPIRE 2002. LNCS, vol. 2476, pp. 1–10. Springer, Heidelberg (2002). https://doi.org/10.1007/3-540-45735-6_1
57. Clement, C.B., Bierbaum, M., O'Keeffe, K.P., Alemi, A.A.: On the use of arxiv as a dataset, arXiv preprint arXiv:1905.00075 (2019)
58. Pinter, Y., Jacobs, C.L., Bittker, M.: Nytwit: a dataset of novel words in the New York times, arXiv preprint arXiv:2003.03444 (2020)
59. Yan, Z., Tang, X.: Exploring evolution of public opinions on tianya club using dynamic topic models. J. Syst. Sci. Inf. **8**(4), 309–324 (2020)

60. Newman, D., Lau, J.H., Grieser, K., Baldwin, T.: Automatic evaluation of topic coherence. In: Human Language Technologies: The 2010 Annual Conference of the North American chapter of the Association for Computational Linguistics, pp. 100–108 (2010)
61. Röder, M., Both, A., Hinneburg, A.: Exploring the space of topic coherence measures. In: Proceedings of the Eighth ACM International Conference on Web Search and Data Mining, pp. 399–408 (2015)
62. Rahimi, H., Hoover, J.L., Mimno, D., Naacke, H., Constantin, C., Amann, B.: Contextualized topic coherence metrics (2023)
63. Hashimoto, T., Shepard, D.L., Kuboyama, T., Shin, K., Kobayashi, R., Uno, T.: Analyzing temporal patterns of topic diversity using graph clustering. J. Supercomput. **77**, 4375–4388 (2021)
64. Bouma, G.: Normalized (pointwise) mutual information in collocation extraction. Proc. GSCL **30**, 31–40 (2009)
65. Reimers, N., Gurevych, I.: Sentence-bert: sentence embeddings using siamese bert-networks, arXiv preprint arXiv:1908.10084 (2019)
66. Face, H.: Hugging face (2021). https://huggingface.co. Accessed 01 Feb 2023
67. Anderson, A., Jurafsky, D., McFarland, D.: Towards a computational history of the ACL: 1980-2008. In: Proceedings of the ACL-2012 Special Workshop on Rediscovering 50 Years of Discoveries, pp. 13–21 (2012)

A Data-Driven Model Selection Approach to Spatio-Temporal Prediction

Rocío Zorrilla[1(✉)], Eduardo Ogasawara[2], Patrick Valduriez[3], and Fábio Porto[1]

[1] Laboratório Nacional de Computação Científica – LNCC, Petrópolis, Brazil
romizc@lncc.br
[2] Centro Federal de Educação Tecnológica Celso Sukow da Fonseca – CEFET-RJ, Rio de Janeiro, Brazil
[3] INRIA & LIRMM, Montpellier, France
http://www.lncc.br

Abstract. Spatio-temporal Predictive Queries encompass a spatio temporal constraint, defining a region, a target variable, and an evaluation metric. The output of such queries presents the future values for the target variable computed by predictive models at each point of the spatio-temporal region. Unfortunately, especially for large spatio-temporal domains with millions of points, training temporal models at each spatial domain point is prohibitive. In this work, we propose a data-driven approach for selecting pre-trained temporal models to be applied at each query point. The chosen approach applies a model to a point according to the training and input time series similarity. The approach avoids training a different model for each domain point, saving model training time. Moreover, it provides a technique to decide on the best-trained model to be applied to a point for prediction. In order to assess the applicability of the proposed strategy, we evaluate a case study for temperature forecasting using historical data and auto-regressive models. Computational experiments show that the proposed approach, compared to the baseline, achieves equivalent predictive performance using a composition of pre-trained models at a fraction of the total computational cost.

Keywords: Spatio-temporal · Time-series · Predictive models

1 Introduction

Successfully predicting the behavior of spatio-temporal phenomena based on past observations is essential for a wide range of scientific studies and real-life applications like precipitation nowcasting [32], and climate alert systems [21]. In support of these applications, traditional data processing and time series analysis

---

The authors thanks CAPES, CNPq, and FAPERJ for partially supporting the paper. This work is developed in the context of the HPDaSc INRIA-Brazil Associated Team.

approaches generate predictive models that aim for predictive accuracy at the cost of high execution time and utilization of computational resources [12,35].

More recently, a new class of systems, known as prediction serving systems, has emerged to support trained models scheduling warranting performance and run-time efficiency [10,18,25]. Inspired by the tradition of database systems, predictive serving systems are expected to support prediction requests through a declarative query interface [5]. For spatio-temporal phenomena, the focus of this paper, expressing a predictive query, involves specifying spatio-temporal constraints that define a region, a target variable whose values are to be inferred, and an evaluation metric for the performance of the predictive query. The query outcome then exhibits the target variable's future values on the specified region, computed by predictive models that meet the metric evaluation threshold.

However, we argue that building a query plan to answer a spatio-temporal predictive query is hard from several perspectives. Among them, we are interested in the model selection and allocation problem: for a given spatio-temporal query region, a serving system must automatically build an appropriate plan that chooses between training models or pick pre-trained models for each query region spatial position such that their composition meets the specified performance constraints and covers the requested spatial area for prediction.

In practice, for large spatial domains, such as the Brazilian territory in a weather forecast service, it is not feasible to hold pre-trained models for each possible point of interest. Complementarily, training models at run-time may not be feasible under stringent elapsed-time query constraints, such as in nowcasting applications [26].

This work proposes reusing pre-trained models built in a reduced set of spatio-temporal points, that probably fall outside the query spatio-temporal region, in order to answer predictive queries. By using pre-trained models, we shall meet the real-time prediction execution constraints. However, as the models may have been trained outside the query region, the procedure shall guarantee that the prediction error produced by the composition of models is minimized.

We adopt a data-driven approach to guide the model selection problem. Considering the availability of historical data, the approach pre-processes the data by grouping sequences of the domain using a shape-based similarity measure, which only considers the temporal dimension. The approach trains time series models at each group's representatives sequence. It uses sequence shape similarity between points in the query region to identify candidate models. Finally, it uses a model recommendation strategy to indicate the ones that meet the metric evaluation criteria.

Our experiments explore the robustness of the domain partitioning and the predictive performance of the proposed model composition used to answer spatio-temporal predictive queries. Results indicate comparable predictive quality using a model composition based on cluster representatives, with a fraction of the computational cost. Moreover, our experiments show that a single clustering strategy, with a fixed number of partitions, may not fully reflect the spatial variations of time series shape throughout the data domain. We adopt a time series classification approach, using a deep learning model, to further improve the model selection.

The remaining of this paper is structured as follows. In Sect. 2, we describe the problem formulation; in Sect. 3, we introduce our proposal to tackle the problem described; in Sect. 4, we show the experimental results; in Sect. 5 we discuss related works; and finally, conclusions and future works are given in Sect. 6.

2 Problem Formulation

Let $\mathcal{D} = \{((x,y), s), \text{ with } (x,y) \in \mathbb{R}^2 \text{ and } s = (s_1, s_2, \ldots, s_T) \text{ denotes a univariate time series (u.t.s) with } T \text{ time units }\}$, $\mathcal{D}$ represents a spatio-temporal domain. Let $\mathcal{G} = \{g_1, g_2, \ldots\}$ be a set of predictive models, based on forecasting techniques, that were trained with different univariate time series $s \in \mathcal{D}$. Each model $g \in \mathcal{G}$ is represented as:

$$g = \langle s, A, \mathbf{p}, E_g, \Sigma_g \rangle, \tag{1}$$

where:

- s: input sequence (time series) divided in training, validation and test subsequences,
- A: forecasting technique,
- $\mathbf{p}$: parameters for the forecast technique,
- E_g: in-sample error [13],
- Σ_g: implementation/execution quality metrics.

We use $g(s, t_p, t_f) = (s_{T+1}, \ldots, s_{T+t_f})$ to represent a forecast of t_f time units of s, indicating that t_p time units were used as validation time series to compute E_g. In this context, we are interested in processing a spatio-temporal predictive query (STPQ) Q:

$$Q = \langle R, t_p, t_f, Q_m \rangle, \tag{2}$$

where:

- R: represents the spatial region of interest,
- t_p: $\{s_{T-t_p-1}, \ldots, s_T\}$ validation time units,
- t_f: $\{s_{T+1}, \ldots, s_{T+t_f}\}$ forecast time units ($t_f \geq 1$),
- Q_m: evaluation metric for the predictive output.

We assume $\langle MSE\{E_g; s \in R\}, t_{train}, t_{eval}\rangle$ as an evaluation metric, bounded by Q_m. Thus, we focus on providing an efficient solution to selecting pre-trained models to compose an answer to a STPQ. This process can be integrated into a more general query processing framework outside the scope of this work. Moreover, $\mathcal{D}$ is a dataset that is directly processed in *raw* by queries in database systems like [31].

3 Our Proposal

Given the problem formulation described in Sect. 2, a possible solution could be to pre-train a predictive model for each time series in $\mathcal{D}$. It is sub-optimal as many points would never be queried, and as the time series change, the models need to be re-trained. Another option would be to train models at the query region points in run-time, severely impacting the query response time.

In this paper, we propose a data-driven model selection approach that focuses on grouping historical data representing the behavior of the target variable in the domain. We argue that, by considering only a set of models generated over a time series representative, which generalizes the shape similarity (variations according to the temporal dimension) of other time series in the domain, it is possible to preserve a predictive quality comparable to the baseline approach of using a model for every time series We could then process an STPQ efficiently while maintaining a low error margin.

In our approach, the domain $\mathcal{D}$ comprises univariate time series and their (lat, lon) positions. However, when building or training models for each time series, we do not include the spatial positions as features. This effectively decouples the spatial component from the domain, allowing us to identify and cluster similar time series based purely on their temporal evolution.

Therefore, in Eq. (1), the focus is on the temporal characteristics of the data:

$$g = \langle s, A, p, E_g, \Sigma_g \rangle$$

For spatio-temporal predictive queries, as shown in Eq. (2), spatial information is incorporated separately to apply the time series models across various locations:

$$Q = \langle R, t_p, t_f, Q_m \rangle$$

Here, R represents the spatial region of interest, guiding the use of the predictive models without altering the underlying time series analysis.

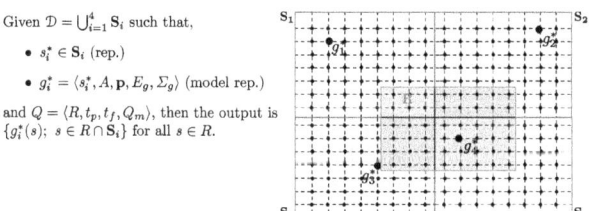

Fig. 1. Our Model Composition Approach

To illustrate our proposal, Fig. 1 shows a domain that has been partitioned into four groups $S = \{S_1, S_2, S_3, S_4\}$. Here, the query region R has 35 univariate time series and intersects with the four groups. Within our proposal, we only need to train four models to process the STPQ. Note that three models were

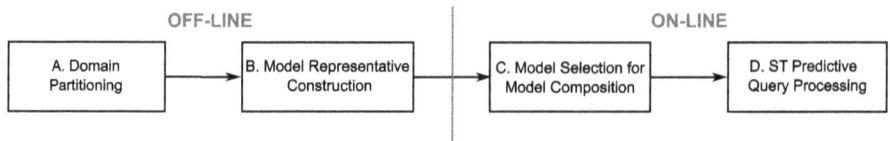

Fig. 2. A two phase query processing approach

trained outside of R. The approach is divided into two phases, offline and online (Fig. 2). The offline phase comprises two steps: (A) the domain partitioning, based on time series clustering techniques; (B) the construction of predictive models at each group time series representative. The online phase is applied when processing a spatio-temporal predictive query. It consists of: (C) a process to select a set of pre-trained representative models, to schedule and run them; (D) an approach to compose the query output using the forecasts of models allocated to every query region point.

The offline phase is also responsible for storing the domain partitioning and the pre-trained models for later retrieval in the online phase. As a result, we can reduce the computational workload and execution elapsed-time if we were to train a model on each point of a query region in run time.

3.1 Domain Partitioning

This step aims to: partition the domain into groups with time series with high shape similarity among themselves; and find a representative. By using k-medoids as a clustering algorithm, each group found can minimize its local dissimilarity and be represented by a medoid that corresponds to an existing time series in the dataset [1]. In this paper, the number of groups k is chosen to produce k corresponding predictive models that produce accurate forecasts for similar time series

Usually, for k-medoids, the choice of k should strike a balance between minimizing the computational cost in using few representatives while maximizing the accuracy when assigning each time series data to its cluster. In the context of the off-line partitioning step, we are not interested in reducing the computational cost. Instead, we want to find the corresponding predictive models that produce accurate forecasts for similar time series (Fig. 3).

The k-medoids algorithm requires the user to specify k. When using a clustering technique for high volumes of data and low variability of the data values throughout neighbor points, it is difficult to determine the optimal number of groups [19]. We consider three methods to find an optimal value for k: the elbow method [1], silhouette index [27] and a fitting of the WSS curve by using a smooth cubic spline [4].

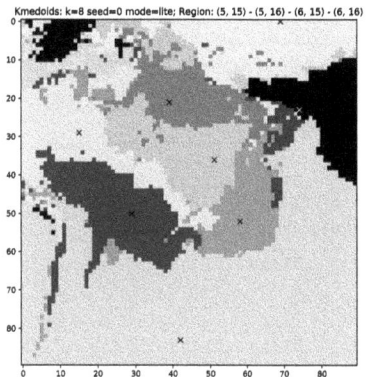 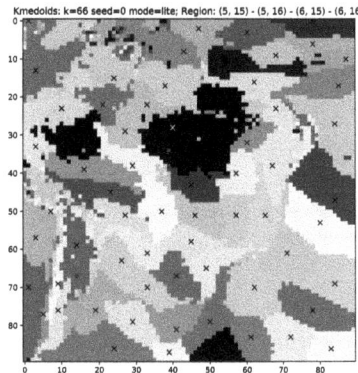

Fig. 3. Groups obtained with k-Medoids using $k = 8$ (left) and $k = 66$ (right). Corresponding medoids are marked with a '×'.

3.2 Model Representative Construction

In order to answer STPQ with acceptable predictive error and reasonable query evaluation time, we consider using a model trained at each medoid. We refer to these k models as representative models and are computed during the offline phase as follows. Let's assume a medoid series has size T. We first train a predictive model using $(T - t_p)$ time units and validate it with the immediate sequence of t_p time units, to compute the forecast error E_g. This model is then re-trained, including the t_p sub-sequence of validation, becoming the representative model that can be used to make predictions of t_f time units for all time series in its group that fall within a particular query region.

3.3 Model Selection for Model Composition

For the scope of this work, we define "Model Composition" as the subset of predictive models that can compute the forecast value of each element in a region of interest on the domain with increased accuracy. The justification to implement this step is based on the intrinsic properties of the spatio-temporal data: the consistency and auto-correlation on nearby points in the domain makes difficult the task of finding an 'optimal' number of groups (k). Even when we consider the elements only in the temporal dimension, this difficulty persists [2]. Within this step, our hypothesis consists in assuming that, if the representative predictive models manage to adequately predict a group of elements with similar shape patterns, then these models will allow us to obtain a prediction for a region of interest of the domain, based on limited information about its past. In order to find this model composition, we consider a model selection process based on the following strategies:

- Naive Approach: for each time series s_j in each group, we train its model g_j and calculate the corresponding forecast error. We consider this the baseline

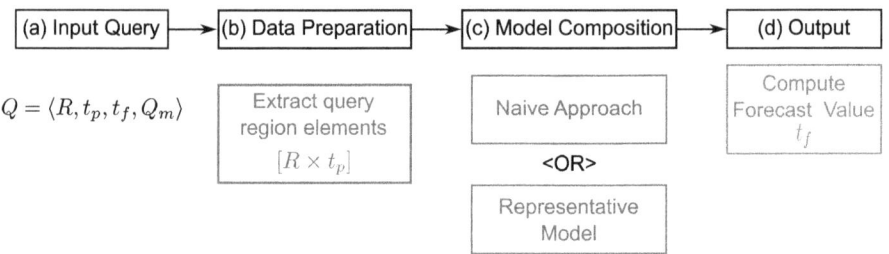

Fig. 4. On-Line STPQ processing.

strategy as it generates as many models as there are time series in the region, and requires a high computational cost.
– Representative Models: we propose that, given the time series representative in each group, we train its corresponding model in order to predict future values for each element in the group and evaluate a corresponding generalization error.

3.4 Spatio-Temporal Predictive Query Processing

The online phase is depicted in Fig. 4, and described as follows:

(a) The query region R and the time units t_p (past) and t_f (future) are parsed from the input query.
(b) A $[R \times t_p]$ spatio-temporal sub-region is extracted from the original dataset, associating a time series of t_p time units for each point in R.
(c) A model composition is created using data about the domain partitioning from the offline phase. Algorithm 1 considers two strategies for model selection: (i) train a predictive model on each point in R, and (ii) intersect the query region R with the groups to find the representatives for every time series and load the pre-trained models.
(d) With the model composition of the previous step, the requested forecast for the t_f steps for each time series in R is computed using its corresponding representative model. Here, we highlight that the same model can generate different forecasts for different time series, provided that the time series undergo a data transformation (e.g., normalization). The forecast is produced by the inverse transformation of the model output.

The online procedure can also be represented by Algorithm 2. As input, the procedure takes the domain, the query parameters, and the model selection strategy. Then, for each element in the query region, the model composition obtained indicates which model performs the forecast, and the known in-sample error of the model is attributed.

Algorithm 1. Apply a Model Selection Strategy

```
 1: function SELECT_MODEL_COMPOSITION(D, selection_id, t_p)
 2:     model_comp ← ⊥
        /* Model Composition with Naive Approach */
 3:     if is_naive_selection(selection_id) then
        /* Let model at each element predict its own element */
 4:         model_comp ← load_trained_models_each(D, t_p)
 5:     end if
        /* Model Composition with Representative Models */
 6:     if is_representative_selection(selection_id) then
        /* User needs to supply value for k of partitioning scheme */
 7:         k ← get_k_for_request(selection_id)
        /* Retrieve previously trained models at each representative */
 8:         (medoids_with_models, D_part) ← load_models_at_medoids(D, k, t_p)
 9:         for m ∈ medoids_with_models do
        /* Retrieve the elements associated to current representative */
10:             cluster ← elements_represented_by(m, D_part)
        /* Let model at current representative predict these elements */
11:             model_comp ← set_predictor(cluster, m, model_comp)
12:         end for
13:     end if
14:     Return model_comp
15: end function
```

Algorithm 2. Process Online Predictive Query

```
 1: function PROCESSQUERY(D, R, t_p, t_f, selection_id)
        /* Obtain a model composition, also load available models /*
 2:     model_comp ← select_model_composition(D, selection_id)
        /* Extract t_p past time units for region R /*
 3:     region_data ← extract_region(D, R, t_p)
 4:     query_out ← ⊥
 5:     for element ∈ region_data do
        /* model composition determines representative (medoid) to use /*
 6:         representative ← find_repr(model_comp, element)
        /* representative has trained model, do forecast of t_f steps /*
 7:         forecast ← predict(representative.model, element, t_f)
        /* annotate the current element with forecast series and known error /*
 8:         error ← representative.error
 9:         annotate(element, forecast, error)
        /* the query result has a set of the annotated elements in R /*
10:         query_result ← add_element(element, query_out)
11:     end for
        /* Compute the MSE of the errors for a single error metric over R /*
12:     error_mse ← combine_errors_mse(query_out)
13:     annotate(query_out, error_mse)
        /* Output is the forecast and error at each element of R, as well as the MSE /*
14:     Return query_out
15: end function
```

4 Experiments and Results

In this Section, we describe the experimental validation of the methodology presented, following the steps: the domain partitioning, the predictive quality of the representative models, the model composition and the query performance. We show how each step is applied to the use case of temperature forecasting, with the corresponding presentation and analysis of the results of each step.

Experimental dataset. We use a subset of the Climate Forecast System Reanalysis (CFSR) dataset, which contains four daily air temperature observations from January 1979 to December 2015 covering the space between

8N-54S latitude and 80W-25W longitude [29]. We subset this data to include one year of readings in the Brazilian territory, then transform each time series into the tuple (latitude, longitude, daily average temperature values), with dimensions (90, 90, 365).

Computational environment. We use a Dell PowerEdge R730 server with 2 Intel Xeon E5-2690 v3 2.60GHz CPUs, 768GB of RAM, and running Linux CentOS 7.7. for all experiments.

4.1 Domain Partitioning Evaluation

We implemented k-medoids using the Dynamic Time Warping similarity measure [30], and compared with a regular partitioning technique (baseline) based solely on the geometry of the domain (k rectangles). The representative time series for each technique are the medoid and centroid, respectively. Computing k-medoids requires pairwise distances, which can be calculated beforehand as a 2-d matrix. In our proposal, we perform this expensive computational process only once, for a quick retrieval later.

For each of the two partition techniques, we vary the number of groups from $k = 2$ up to $k = 150$ with a stride of two and calculate the Within-cluster Sum of Squares (WSS) for each value of k. For k-medoids, we obtain a monotonically decreasing trend for the WSS curve. This makes the choice for an optimal k difficult, a known problem for high volumes of data with low variability throughout neighbor points [19].

In Fig. 5, it is possible to observe the decreasing behavior of the WSS curve for higher values of k.

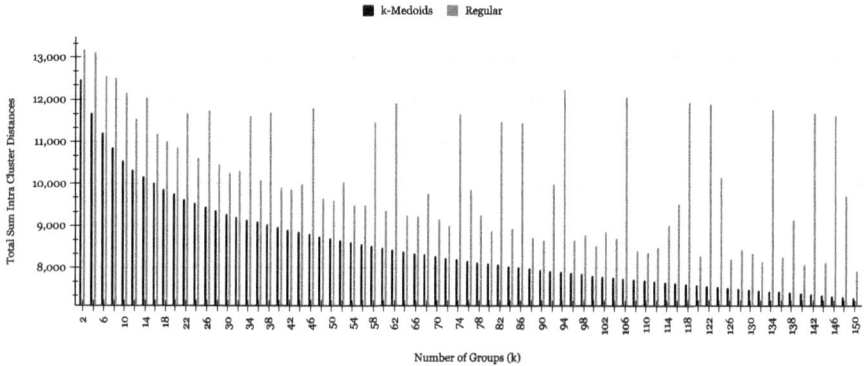

Fig. 5. Total Within cluster Sum of Squares of k-Medoids and Regular Partitioning Techniques.

It is particularly important for our problem, where there is low variation in the spatial data distribution of the different time series The Table 1 summarizes the findings of applying the available methods to find optimal values for k (Sect. 3.1). The monotonous trend of k-medoids allowed for calculating the minimum value for the second derivative, by fitting the values of the WSS using a cubic smooth spline. We argue that this method is more appropriate for our dataset, as it highlights the decreasing trend in the intra-cluster cost as k increased. It was possible because the splines smoothed the small variations that were preventing the other methods from finding a higher value for k. This value gives us a reasonable number of representatives to use in the next phases of the methodology. Also, since we applied two other evaluation methods for selecting k that produced two other domain partitioning schemes, we can also consider these groups when evaluating compositions relevant to our proposed model selection approach.

Table 1. Methods to find the optimal value for k.

Method	Optimal k
Elbow	4
Silhouette	8
Cubic spline for WSS	66

4.2 Predictive Quality of Models at Representatives

Here, we are interested in evaluating the accuracy of the forecast values computed on the test sub-sequence (t_f) by comparing them against the observational values available. In this work, we consider the Symmetric Mean Absolute Percentage Error (sMAPE) for forecast error evaluation and the Mean Squared Error (MSE) [15] for accumulated forecast (Fig. 6).

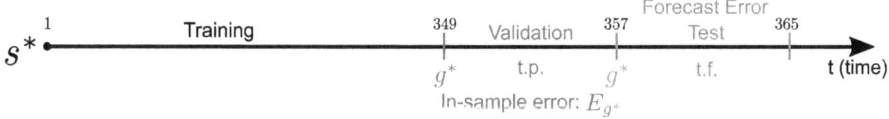

Fig. 6. Splitting a Sequence to train and test a model.

In order to assess the predictive quality of a model on a representative time series, we train the corresponding k models in a domain partitioning and evaluate the following metrics:

- sMAPE: Forecast error of the representative model when forecasting each time series in its respective group.
- MSE: Accumulated forecast error (sMAPE) by combining the previous forecasts within each group.

In this section, we evaluate the predictive quality of Auto-Regressive Integrated Moving Average (ARIMA) models [3]. These models fit the description in Sect. 2 and offer a good trade-off between predictive accuracy and computational cost [21]. We leverage auto.ARIMA [14] implementation to choose optimal ARIMA parameters.

Evaluation of sMAPE Forecast Error. Considering the domain partitioning with $k = 8$, we have eight groups with 1013 ± 617 time series on average, yielding eight representative models. In order to explore the relationship between the intra-cluster similarity and forecast error, we gather the forecast errors within each group and generate scatter plots diagrams, with the Dynamic Time Warping distance of each sequence to its medoid in the x-axis and the sMAPE metric in the y-axis. We found that, for the group index zero (Fig. 7), the maximum sMAPE value was the lowest among the eight groups. Conversely, for the group index four (Fig. 8), the maximum sMAPE value was the highest.

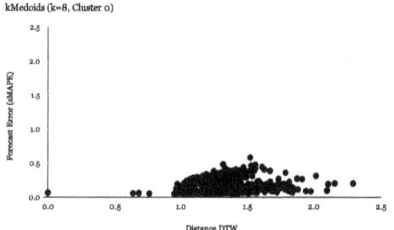

Fig. 7. Forecast Error $= 0.159 \pm 0.073$.   **Fig. 8.** Forecast Error $= 0.718 \pm 0.347$.

We observe that there is not a clear correlation between the Dynamic Time Warping distances and the forecast error. If we consider all the representatives, then as k increases, there is a tendency to obtain groups with more similarity between their elements (lower Dynamic Time Warping distance) and also the predictions tend to be more accurate. An additional important observation here is that lower values of k (8, 66) can produce some representatives that offer better predictions than, for example, the 'worst' (highest forecast error) representatives of the partitioning scheme with $k = 132$. Both these observations indicate that different spatial areas may need more precise partitioning than others.

Evaluation of MSE Forecast Error. Here we are interested in evaluating the MSE metric computed when forecasting an entire group of domain partitioning. We compare the following approaches:

– ARIMA or Naive Approach (Baseline): We train a model using a time series and calculate the corresponding forecast error for every time series in each group. Then for each group, we compute its corresponding MSE value.

– Representative Models (Proposal): given the k corresponding models for the representatives in a domain partitioning, we use its representative model to forecast future values; finally, we compute the accumulated MSE values.

Considering the domain partitioning with $k = 8$, Table 2 highlights the results of the MSE evaluation. The columns are as follows: (1) cluster/group ID; (2) elapsed time to train models for all the time series in the group (naive approach); (3) elapsed time to forecast t_f future units for the time series in the group (naive approach); (4) accumulated MSE value for the Naive Approach; (5) accumulated MSE value for the Representative Models; (6) percentage change of the MSE values between the approaches.

Col. 1. Group or Cluster Id.
Col. 2. Elapsed time to train models for all the time series in the group (naive approach).
Col. 3. Elapsed time to forecast t_f future units for the time series in the group (naive approach).
Col. 4. Accumulated MSE value for the Naive Approach.
Col. 5. Accumulated MSE value for the Representative Models.
Col. 6. Percentage change of the MSE values between the approaches.

Table 2. Forecast Error Analysis with $k = 8$ and $t_f = 8$.

cid	T. Train.(s)	T. For. (s)	ARIMA	Repr. Models	Δ (%)
0	2041.469	1.069	0.170	0.185	8.82
1	3447.608	1.299	0.689	0.926	34.38
2	2011.441	0.880	0.581	0.678	16.70
3	2685.912	1.238	0.413	0.492	19.13
4	14542.318	5.727	0.785	0.838	6.75
5	3231.718	1.375	0.407	0.437	7.37
6	1930.740	0.957	0.157	0.203	29.30
7	1811.335	0.853	0.388	0.551	42.01

We observe that the MSE of the Representative Models varies significantly between groups and is consistently larger than the MSE of the Naive Approach, by 6.75% to 42.01%. Moreover, we find that 76% of the domain time series can be predicted using only five models with a forecast error incremented by at most 20% of the Naive Approach, which would consider 8100 different models for the same predictions. These results support our hypothesis that when considering more compact groups, each representative generalizes its elements better, and this generalization can be extended to the predictive quality.

Elapsed Time for Training, Validation and Forecast. An additional aspect in the evaluation of the Representative Models is the computational cost for training and forecasting. According to Table 2, the total time for training the models over all the time series in the Naive Approach is about 31500 s (8.75 h). Thus, the average training time of an ARIMA model using a time series with 349 time units is $31500/8100 \approx 3.9$ s. In our proposal, we consider train and re-train models for k representative time series Thus, the total training time for a given partitioning can be estimated as $k \times (2 \times 3.9)$ seconds, about a minute for the domain partitioning with $k = 8$.

The results in this section support the hypothesis that: (1) the data distribution variation observed in different regions of the domain would point to a strategy based on multiple partitioning criteria; (2) by using model representatives, we can significantly reduce the model training cost while keeping acceptable forecast errors. Additionally, experiments in this section were repeated for all values of k considered in Sect. 4.1, and we found that $k = 132$ minimized the MSE metric. For these reasons, we will consider $k = \{8, 66, 132\}$ for multiple domain partitioning criteria.

4.3 Processing Spatio-Temporal Predictive Queries

Our proposed online phase (See Fig. 2) comprises two steps: (D) Model Selection for Model Composition and (E) STPQ Processing. Here, we evaluate the predictive quality of a Model Composition over a region of interest R when processing an STPQ.

Model Composition Evaluation. To assess the predictive quality of a Model Composition, we consider multiple domain partitioning criteria and a Model Selection approach to be applied on query regions of fixed size $R = [10 \times 10]$ distributed uniformly over the domain. We consider these approaches for Model Selection:

- Naive Selection: For each time series in R, we select its pre-trained ARIMA model. Here, we introduce kNN as an additional predictive model for comparisons of forecast accuracy, so we obtain two sets of experimental results.
- Selection of Representative Models: For each time series in R, we determine its corresponding group and select the pre-trained ARIMA Representative Model.

The predictive quality of the Model Composition forecasts is evaluated using the accumulated MSE over the query region R. These results are summarized in Table 3: the first column corresponds to the Model Composition using Naive Selection; the following three columns represent the Model Composition formed by the Selection of Representative Models for a domain partitioning, varying $k = \{8, 66, 132\}$. We present the mean and variance of the MSE for the 81 query regions.

Table 3. MSE Forecast Error Summary.

ARIMA	kNN	k-Medoids		
		$k = 8$	$k = 66$	$k = 132$
0.38 ± 0.61	0.62 ± 0.91	0.48 ± 0.59	0.47 ± 0.86	0.39 ± 0.62

We use color maps to help the visualization of the MSE over different regions of the domain. Figure 9 and Fig. 10 correspond to $k = 66$ and $k = 132$, respectively. Each color map shows the relative magnitude of the values, with a dark blue for the highest forecast error and a constant palette throughout figures.

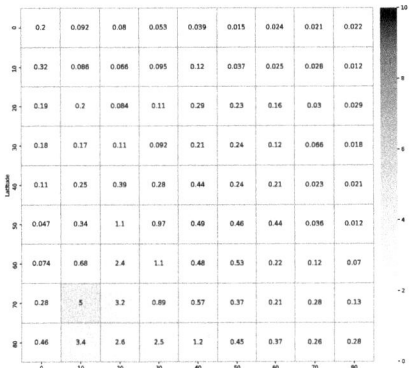

Fig. 9. Model Composition with Representatives ($k = 66$).

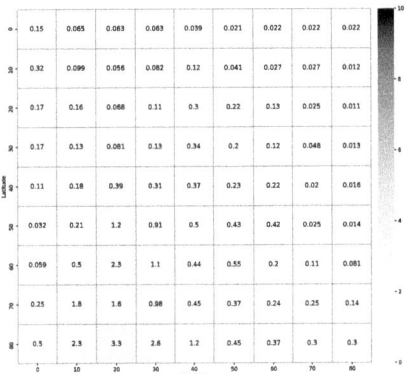

Fig. 10. Model Composition with Representatives ($k = 132$).

Experimentally, we find that a spatial region near the bottom left results in larger forecast errors. Even there, using k (8, 66) may yield better results than $k = 132$ for some slices. This finding triggered the development and evaluation that follows next.

Classifier for Model Selection. This section proposes a Model Selection approach that leverages the predictive quality variation of the Representative Models in domain partitioning. Here, the intuition is that by applying multiple partitioning to a domain, each time series would be mapped to a set of groups. Conversely, each domain sequence would be associated with a set of model representatives, and so the question is which one to pick.

We extend the problem formulation presented in Sect. 2. Consider two domain partitioning criteria $\mathcal{D} = \cup_{i=1}^{m} \mathbf{P_i}$ and $\mathcal{D} = \cup_{j=1}^{n} \mathbf{Q_j}$, where $m \neq n$; the set of representatives on the partitioning considered is $\mathcal{R} = \{p_i, \ldots, p_m, q_1, \ldots, q_n\}$, and $\mathcal{G}_{(\mathcal{R})}$ the set of their predictive models. Then, for a given $s \in \mathcal{D}$:

$$\exists \hat{s} \in \mathcal{R}, \text{ such that, } \min_{\substack{\hat{s} \in \mathcal{R} \\ s \in \mathcal{D}}} d_{DTW}(\hat{s}, s). \tag{3}$$

We formulate the model selection proposal as a univariate time series (univariate time series) classification problem: Given an unlabeled univariate time series of t_p time units, assign it to one or more predefined classes. From (3), we are able to generate the Time Series Classification Dataset as $TSCD = \{(s_1, y_1), \ldots, (s_N, y_N)\}$ as a collection of pairs (s_i, y_i) where s_i is a u.t.s with y_i as its corresponding one-hot label vector of the labels for its class [11].

In our context, each of these classes represents one of the available domain partitioning criteria. Considering $k = \{8, 66, 132\}$, we obtain 183 classes in total, after accounting for medoid repetition. In order to work with a balanced dataset, we extract for the $TSCD$ approximately 30 samples per class [8]. We consider 5000 samples, divided in the percentages 60/20/20 for training, validation, and test, respectively.

Considering the sequential aspect of time series data requires algorithms that can harness this temporal property to select a class label. In this work, we consider a classifier based on Neural Network models. After considering non-hybrid approaches that provided inferior classification accuracy [16], we opted for the hybrid architecture 1D Convolutional Neural Network – Long-Short Term Memory (1DCNN-LSTM) [7,34]. We considered variations for parameters such as learning rate and batch size, that affect the training time and how fast we achieve convergence in the validation loss function. From now on, we will consider only the last model in Table 4 (CNN1D-LSTM(2)) that presented the higher accuracy.

Table 4. Models' metrics on Test Set.

Model	Layers	Accuracy	Loss
CNN1D-LSTM(1)	6	57.740	2.673
CNN1D-LSTM(2)	6	64.759	1.865

Evaluation of the Classifier for Model Selection. After training the Classifier presented in the previous section, we repeat the same experiments from Sect. 4.3 using the classifier as a Model Selection approach. For each time series in R, the classifier receives a time series of length t_p as input. As output, we obtain a Representative Label that corresponds to one of the Representative Models. With this model selection process, we repeat the forecast error analysis from Sect. 4.3.

Table 5. MSE Forecast Error Summary including the Classifier.

ARIMA	kNN	k-Medoids			Classifier
		$k = 8$	$k = 66$	$k = 132$	
0.38 ± 0.61	0.62 ± 0.91	0.48 ± 0.59	0.47 ± 0.86	0.39 ± 0.62	0.70 ± 0.81

The experimental results are summarized in Table 5: it extends the Table 3 with the last column (highlighted) representing the Classifier for Model Selection.

We show the colormap for the MSE of the forecast errors computed in different regions of the domain using the Classifier for model composition in Fig. 11. For comparison, we also add the colormap for the ARIMA (baseline) Approach (Fig. 12).

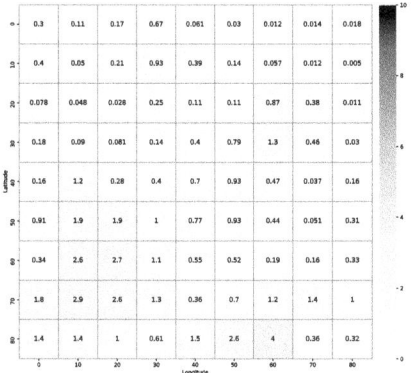

Fig. 11. Forecast Error with Model Composition by Classifier.

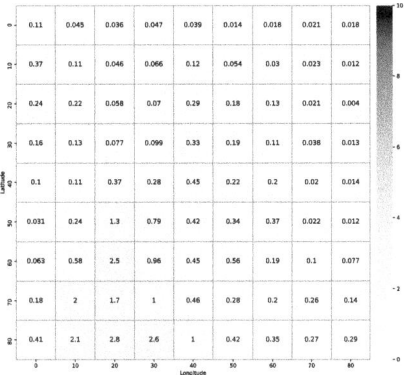

Fig. 12. Model Composition with ARIMA, naive approach.

We observed that the Classifier generates a composition with predictive quality comparable to the Naive Approach in some areas. Moreover, the classifier quality is reflected in a few regions of the domain that exhibit a smaller forecast error than when using the composition based on Selection of Representative Models directly. However, the opposite is true for other regions, this can be explained by the limited knowledge of the classifier about the time series, as it receives time series of t_p time units.

Finally, we compare the execution of an STPQ using the proposed Model Composition, with the Naive Selection based on ARIMA models and kNN regressions for univariate time series, over different query region sizes. Results are shown in Table 6, it is similar to Table 5 but with the query regions. We observe that, for the majority of the query regions considered, the forecast error of the Classifier for Model Selection is closer to the ARIMA Naive Selection.

Table 6. MSE Forecast Error for Spatio-Temporal Queries in the domain $\mathcal{D}$.

Query Region	ARIMA	kNN	k–Medoids			Classifier
			$k=8$	$k=66$	$k=132$	
$[0,20] \times [0,20]$	0.158	0.209	0.089	0.174	0.160	0.190
$[20,40] \times [35,55]$	0.203	0.258	0.335	0.199	0.230	0.330
$[50,70] \times [60,80]$	0.170	0.145	0.584	0.203	0.188	0.274
$[15,35] \times [65,85]$	0.034	0.067	0.063	0.045	0.038	0.093
$[20,50] \times [50,80]$	0.122	0.210	0.203	0.147	0.135	0.202
$[15,45] \times [20,50]$	0.156	0.198	0.262	0.155	0.168	0.281
$[40,55] \times [20,40]$	0.483	0.707	0.707	0.530	0.541	0.618
$[65,80] \times [50,70]$	0.248	0.190	0.470	0.302	0.308	0.343
$[30,60] \times [5,20]$	0.137	0.208	0.353	0.205	0.147	0.391
$[10,40] \times [55,70]$	0.095	0.226	0.139	0.111	0.098	0.135

5 Related Works

In this work, we integrate tools designed for two types of knowledge fields: (i) time series classification and (ii) processing spatio-temporal predictive queries. The former gained attention in the last decade due to the accelerated advancement of deep learning techniques, many are discussed in a thesis aimed at deep learning for TSC [16], and the site http://www.timeseriesclassification.com, in efforts to reunite dataset and research papers on this evolving topic.

For time series clustering, the use of k-medoids with Dynamic Time Warping as similarity measure was used with success in several applications [22,28]. In our work, we validate k–Medoids as an appropriate algorithm for our dataset, but we shifted our focus away from the Euclidean distance in favor of Dynamic Time Warping; the former failed to capture temporal misalignments.

Common uses for spatio-temporal predictive queries in spatio-temporal data are predictive analytics to answer complex questions involving missing or future values, correlations, and trends, which can be used to identify opportunities or threats [10,25]. The predictive functionality can help build introspective services for various resource management and optimization tasks [9].

While we do not aim to propose a full Predictive Serving System [6], it is worth exploring some of these systems to better understand the requirements behind model composition and model selection. The framework Clipper [6] is designed to serve trained models at interactive latency, with two model selection policies based on multi-armed bandit algorithms for a trade-off between accuracy and computation overhead. Rafiki [33] is an inference service based on reinforcement learning that provides an online multi-model selection to compose ensembles.

Regarding massive data processing and model training, in [20] are discussed techniques for dataset characterization in a reduced number of representatives elements, with data-efficient methods to extract representative subsets that generalize the full data. The work focuses on extracting representative subsets for training machine learning models, and developing theoretically rigorous optimization techniques. Finally, DJEnsemble [24] investigates the prediction of spatio-temporal phenomena using deep-learning models. However, instead of our shape-based approach, they partition the domain into tiles based on the statistical properties of the time series in contrast of our shape-based approach.

6 Conclusions and Future Works

The main objective of this work is to develop an approach to make predictions, within some tolerated error margin, about future states of a spatio-temporal region, using carefully selected predictive models that have been trained with limited temporal data. To achieve this, we formulate the problem of model composition to process predictive queries and propose a solution where the model selection is guided by a data-driven approach backed by shape-based domain partitioning. The computational experiments were then designed to evaluate the proposal, considering the case study of temperature forecasting.

Experimentally, we find that the k-medoids method can efficiently group time series in a domain according to the Dynamic Time Warping distance. Also, the resulting medoids generalizes the temporal evolution of their group. Therefore it can be used to train representative models that take a univariate time series as input. Within our proposal, both the domain partitioning (k–medoids) and the construction of Representative Models can be computed and persisted during an offline phase, quickly retrieved during an online phase, significantly reducing the elapsed time for processing predictive queries. In this regard, the choice of k becomes an important factor for the predictive quality, and three techniques to find optimal values of k were explored. We find that the intuitive choice of a large value of k may not always produce the best results: fewer groups may produce more accurate results for some elements of the query region.

The previous result motivated the proposal of a neural network classifier for model selection. In the offline phase, we allow the construction of representative predictive models for multiple partitioning criteria ($k = \{8, 66, 132\}$). For the online phase, the classifier matches the subset (t_p time units) of each univariate time series in the query region to one of the representatives, thus creating the model composition for a given predictive query.

We show that our proposal can process predictive queries with significantly lower response time, while maintaining comparable predictive quality. To evaluate this experimentally, we used sMAPE forecast errors accumulated over query regions with MSE. Results indicate 20% and 45% relative increases for $k = 66$ and the Classifier approach, respectively, with a gain in computational efficiency of two orders of magnitude as a trade-off.

Results from the forecast error analysis support using a time series classifier to leverage potential gains in predictive performance when using multiple partitioning schemes. However, we recognize that the Classifier needs to be improved, e.g., by considering a domain with a larger volume of data and understanding its classification accuracy.

Our proposal opens up several research directions. The calculation of pairwise Dynamic Time Warping distances can be enhanced by grouping time series with an incremental process for the Dynamic Time Warping matrix [23]. For the domain partitioning task, we could consider non-crisp partitioning techniques [17], producing more than one representative for a given element. This work did not focus on forecast time for the online phase as the ARIMA models deliver predictions in milliseconds (see Table 2); however, more complex models would imply significant service times. Therefore, a natural follow-up would include a multi-objective optimization process.

References

1. Aggarwal, C.C., Reddy, C.K.: Data Clustering: Algorithms and Applications, 1st edn. Chapman and Hall/CRC, London (2013)
2. Aghabozorgi, S., Seyed Shirkhorshidi, A., Ying Wah, T.: Time-series clustering - a decade review. Inf. Syst. **53**(C), 16-38 (2015). https://doi.org/10.1016/j.is.2015.04.007
3. Box, G., Jenkins, G.M.: Time Series Analysis: Forecasting and Control. Holden-Day, San Francisco (1976)
4. Burden, R.L., Faires, D.J., Burden, A.M.: Numerical Analysis, 10th edn. CENGAGE Learning, Boston (2016)
5. Crankshaw, D., Gonzalez, J., Bailis, P.: Research for practice: prediction-serving systems. Commun. ACM **61**(8), 45–49 (2018). https://doi.org/10.1145/3190574
6. Crankshaw, D., Wang, X., Zhou, G., Franklin, M.J., Gonzalez, J.E., Stoica, I.: Clipper: a low-latency online prediction serving system. In: 14th USENIX Symposium on Networked Systems Design and Implementation (NSDI 2017), Boston, MA, pp. 613–627. USENIX Association (2017)
7. Du, Q., Gu, W., Zhang, L., Huang, S.L.: Attention-based LSTM-CNNs for time-series classification. In: Proceedings of the 16th ACM Conference on Embedded Networked Sensor Systems, SenSys 2018, pp. 410–411. Association for Computing Machinery, New York (2018). https://doi.org/10.1145/3274783.3275208
8. Du, S.S., Wang, Y., Zhai, X., Balakrishnan, S., Salakhutdinov, R.R., Singh, A.: How many samples are needed to estimate a convolutional neural network? In: Bengio, S., Wallach, H., Larochelle, H., Grauman, K., Cesa-Bianchi, N., Garnett, R. (eds.) Advances in Neural Information Processing Systems, vol. 31. Curran Associates, Inc. (2018). https://proceedings.neurips.cc/paper/2018/file/03c6b06952c750899bb03d998e631860-Paper.pdf
9. Filippo, A.D., Lombardi, M., Milano, M.: Methods for off-line/on-line optimization under uncertainty. In: Proceedings of the Twenty-Seventh International Joint Conference on Artificial Intelligence, IJCAI 2018, pp. 1270–1276. International Joint Conferences on Artificial Intelligence Organization (2018). https://doi.org/10.24963/ijcai.2018/177
10. Ghanta, S., et al.: ML health monitor: taking the pulse of machine learning algorithms in production. In: Zelinski, M.E., Taha, T.M., Howe, J., Awwal, A.A.S., Iftekharuddin, K.M. (eds.) Applications of Machine Learning, vol. 11139, pp. 191–202. International Society for Optics and Photonics, SPIE (2019). https://doi.org/10.1117/12.2529598
11. Gulli, A., Pal, S.: Deep Learning with Keras. Packt Publishing, Birmingham (2017)

12. Hassani, H., Silva, E.S.: Forecasting with big data: a review. Ann. Data Sci. **2**(1), 5–19 (2015). https://doi.org/10.1007/s40745-015-0029-9
13. Hastie, T., Tibshirani, R., Friedman, J.H.: The Elements of Statistical Learning: Data Mining, Inference, and Prediction, 2nd edn. Springer Series in Statistics. Springer, New York (2009)
14. Hyndman, R.J., Khandakar, Y.: Automatic time series forecasting: the forecast package for R. J. Stat. Softw. Articles **27**(3), 1–22 (2008). https://doi.org/10.18637/jss.v027.i03
15. Hyndman, R.J., Koehler, A.B.: Another look at measures of forecast accuracy. Int. J. Forecast. **22**(4), 679–688 (2006). https://doi.org/10.1016/j.ijforecast.2006.03.001
16. Ismail Fawaz, H., Forestier, G., Weber, J., Idoumghar, L., Muller, P.A.: Deep learning for time series classification: a review. Data Min. Knowl. Discov. **33**(4), 917-963 (2019). https://doi.org/10.1007/s10618-019-00619-1
17. Izakian, H., Pedrycz, W., Jamal, I.: Fuzzy clustering of time series data using dynamic time warping distance. Eng. Appl. Artif. Intell. **39**, 235–244 (2015). https://doi.org/10.1016/j.engappai.2014.12.015
18. Lee, Y., Scolari, A., Interlandi, M., Weimer, M., Chun, B.G.: Towards high-performance prediction serving systems. In: NIPS Machine Learning Systems Workshop (2017)
19. Liao, T.W.: Clustering of time series data: a survey. Pattern Recogn. **38**(11), 1857–1874 (2005). https://doi.org/10.1016/j.patcog.2005.01.025
20. Mirzasoleiman, B.: Efficient machine learning from massive datasets (2021). http://web.cs.ucla.edu/~baharan/research.htm
21. Murat, M., Malinowska, I., Gos, M., Krzyszczak, J.: Forecasting daily meteorological time series using ARIMA and regression models. Inter. Agrophys. **32**(2), 253–264 (2018). https://doi.org/10.1515/intag-2017-0007
22. Nakagawa, K., Imamura, M., Yoshida, K.: Stock price prediction using k-medoids clustering with indexing dynamic time warping. Electron. Commun. Japan **102**(2), 3–8 (2019). https://doi.org/10.1002/ecj.12140
23. Oregi, I., Pérez, A., Del Ser, J., Lozano, J.A.: On-line dynamic time warping for streaming time series. In: Ceci, M., Hollmén, J., Todorovski, L., Vens, C., Džeroski, S. (eds.) ECML PKDD 2017. LNCS (LNAI), vol. 10535, pp. 591–605. Springer, Cham (2017). https://doi.org/10.1007/978-3-319-71246-8_36
24. Pereira, R., et al.: DJEnsemble: a cost-based selection and allocation of a disjoint ensemble of spatio-temporal models, pp. 226–231. Association for Computing Machinery, New York (2021). https://doi.org/10.1145/3468791.3468806
25. Polyzotis, N., Roy, S., Whang, S.E., Zinkevich, M.: Data lifecycle challenges in production machine learning: a survey. SIGMOD Rec. **47**(2), 17–28 (2018). https://doi.org/10.1145/3299887.3299891
26. Ravuri, S.V., et al.: Skilful precipitation nowcasting using deep generative models of radar. Nature **597**, 672–677 (2021)
27. Rousseeuw, P.J.: Silhouettes: a graphical aid to the interpretation and validation of cluster analysis. J. Comput. Appl. Math. **20**, 53–65 (1987). https://doi.org/10.1016/0377-0427(87)90125-7
28. Ruiz, L., Pegalajar, M., Arcucci, R., Molina-Solana, M.: A time-series clustering methodology for knowledge extraction in energy consumption data. Expert Syst. Appl. **160**, 113731 (2020). https://doi.org/10.1016/j.eswa.2020.113731
29. Saha, S., et al.: The NCEP climate forecast system reanalysis. Bull. Am. Meteorol. Soc. **91**(8), 1015 – 1058 (2010). https://doi.org/10.1175/2010BAMS3001.1

30. Sakoe, H., Chiba, S.: Dynamic programming algorithm optimization for spoken word recognition. IEEE Trans. Acoust. Speech Signal Process. **26**(1), 43–49 (1978). https://doi.org/10.1109/TASSP.1978.1163055
31. da Silva, A.C., Lustosa, H.L.S., da Silva, D.N.R., Porto, F.A.M., Valduriez, P.: SAVIME: an array DBMS for simulation analysis and ML models prediction. J. Inf. Data Manag. **11**(3) (2020). https://periodicos.ufmg.br/index.php/jidm/article/view/24223
32. Souto, Y.M., Porto, F., de Carvalho Moura, A.M., Bezerra, E.: A spatiotemporal ensemble approach to rainfall forecasting. In: 2018 International Joint Conference on Neural Networks, IJCNN 2018, Rio de Janeiro, Brazil, 8–13 July 2018, pp. 1–8 (2018)
33. Wang, W., et al.: Rafiki: machine learning as an analytics service system. Proc. VLDB Endow. **12**(2), 128–140 (2018). https://doi.org/10.14778/3282495.3282499
34. Xu, G., Ren, T., Chen, Y., Che, W.: A one-dimensional CNN-LSTM model for epileptic seizure recognition using EEG signal analysis. Front. Neurosci. **14**, 1253 (2020). https://doi.org/10.3389/fnins.2020.578126
35. Yang, C., Clarke, K., Shekhar, S., Tao, C.V.: Big spatiotemporal data analytics: a research and innovation frontier. Int. J. Geogr. Inf. Sci. **34**(6), 1075–1088 (2020). https://doi.org/10.1080/13658816.2019.1698743

Optimistic Data Generation for JSON Schema

Lyes Attouche[1](✉) , Mohamed-Amine Baazizi[2] , and Dario Colazzo[3]

[1] Université Paris-Dauphine, PSL Research University, Paris, France
`lyes.attouche@dauphine.eu`
[2] Sorbonne Université, LIP6 UMR 7606, Paris, France
`baazizi@ia.lip6.fr`
[3] Université Paris-Dauphine, PSL Research University, Paris, France
`dario.colazzo@dauphine.eu`

Abstract. JSON Schema is an expressive schema language for describing JSON documents which combines structural assertions with Boolean operators, including negation, and is able to express recursion. In typical situations, JSON Schema is used to validate JSON documents to ensure safety of operations processing these JSONdata. In many other situations, it is important to *generate* JSONdocuments starting from a JSON Schema schema as it allows for testing applications and for assessing interesting properties like scalability of processing engines. While existing work [6,8] addressed the *witness* generation problem, by discarding the `uniqueItems` operator, our goal lifts this restriction and addresses the problem of generating multiple instances from a given schema. Since the witness generation problem, which is equivalent to checking the satisfiability in JSON Schema, is known to be EXPTIME-complete even when `uniqueItems` is absent [10], our solution adopts an *optimistic* approach which trades *completeness* for *soundness* and *efficiency*. From a practical point of view, our approach is very satisfactory since, on most real-life scenarios, it is very efficient and highly precise, as testified by our experiments.

Keywords: JSON · JSON Schema · Data Generation

1 Introduction

JSON Schema is a language for describing families of JSON documents. It is increasingly adopted simply because the number of APIs using JSON as a format for exchanging data explodes. In this context, it is paramount to be able to decide whether a given schema is *satisfiable* and to enable for generating testing data starting from a JSON Schema specification. This ability may serve many applications: for instance, one can simulate and test the behaviour of a program on synthetic data in order to anticipate potential mismatches or performance degradation whose consequences may have a dramatic impact on the reliability of services once the program is put in production. Generating multiple instances of a JSON Schema is also important for schema understanding: in a previous research we have realized that JSON Schema schemas used in practice are rather complex and large. Having the possibility of generating several instances may facilitate for the user the understanding of such schemas, but also speed up operations related to checking schema evolution and implementation of safe applications over the schema.

We have recently worked on approaches to generate a single witness for a JSON schema, under the condition that `uniqueItems` for array description is not used [6,8]. The approach is sound and complete, but currently is not meant to generate a set of witnesses (instances), and for schema featuring particular operators, generation time could be prohibitive (witness generation has been proved to be EXPTIME complete for JSON Schema by discarding `uniqueItems` [10]). This is due to the fact that our previous approach rewrites the schema, before generation, into a novel form of disjunctive normal form that, on the one hand is time expensive to generate, and, on the other hand is expensive to explore. Schema rewriting, also called *preparation*, is necessary in order to get completeness, as JSON Schema has a very high expressive power, and witness generation is a complex task: preparation is necessary in order to deal with mechanisms that are very complex to deal with for generation purposes. So this approach can be considered as a *pessimistic* one, in the sense that it performs complex and preliminary operations, assuming that operators that are very difficult to manipulate are present in the schema. While the approach we presented in [6,8] is also able to deal with schema inclusion checking, another crucial operation in the context of JSON Schema analysis, in the context of data generation our experience has shown that in many cases JSON schemas, even if they remain rather complex to deal with, are not extremely complex, and can be handled according to different, faster, and of course incomplete approaches.

The alternative approach that we present in this paper, and that we dub *optimistic* generation, is based on adopting a much faster schema rewriting before generation, and on generating schema instances in a *greedy* fashion, by identifying an opportune priority order into which interacting statements for array and objects are considered for data generation. As shown by extensive experiments, the resulting approach is fast in the vast majority of schemas, even for schemas featuring complex operators, like interacting patterns for object and array operators, including uniqueItems, which was not considered in previous work and which introduces substantial complications. In particular, in order to deal with uniqueItems we devise a technique which is able to generate $n \geq 1$ instances so as to ensure uniqueness in most cases. So we make an important step towards effective multi instance generation, while our previous approach is focused on one single witness/instance generation [6,8]. In this context, efficiency that we guarantee by means of our optimistic strategy is crucial, especially when generating a high number of instances, which turns out to be particularly useful when one needs to test scalability of JSON processing applications.

The paper is structured as follows. In Sect. 2, we discuss related works, while in Sect. 3, we present the JSON Schema language, first by giving an overview and then by detailing its semantics. In Sect. 4, we introduce the pre-processing phase of our approach, followed by the generation phase in Sect. 5. In Sect. 6, we present experimental results validating the effectiveness of the solution. We draw our conclusions in Sect. 7.

2 Related Work

Static analysis over JSON schema has attracted interest in the last decade, with a particular focus on satisfiability and inclusion checking, as well as single-witness generation, which also allows to solve the two previous problems. Satisfiability has been shown to be EXPTIME-complete for JSON Schema without uniqueItems, and 2EXPTIME when it is included [10]. Unfortunately, as observed in [8], algorithms directly derived by these proofs are of no practical interest, due to execution time blowing up in most cases. In [8] we provided a sound and complete (hence exponential) checking algorithm showing good performances in most of practical cases, and also having the advantage of producing a witness whenever the schema is satisfiable. Differently from the approach we present here, in [8] uniqueItems is not considered, and what is guaranteed is single-witness generation. Still concerning satisfiability and inclusion, the inclusion checker described by Habib et al. [11] which allows for checking that a schema S_1 is contained in S_2, denoted $S_1 \subseteq S_2$ can also be exploited for schema satisfiability since S is satisfiable if and only if $S \nsubseteq S'$, where S' is an empty schema. However, this approach has a limited ability when dealing with negation and recursion, and is not able to generate any schema instance.

With respect to the previous conference version of this work [7], in this extended version we make the following additional contributions. More details are provided concerning main algorithms, in particular by providing their pseudocode as well as illustrative examples. A novel contribution of this paper is generation of multi instances, which was not dealt with in the previous short paper [7]. Finally we present here results on new experimental analysis about unique items and multi-instance generation.

Several open-source tools for example generation exist but very few of them are robust and many are no longer maintained. We selected the most reliable tools by analyzing their limitations and we experimented with them over real-world schema collections that we obtained from [8] and found that only three of them ([2,3] and [1]) have an average correction rate exceeding 50%. We describe their behaviour and major limitations and leave the quantitative analysis to Sect. 6.

Data Generator (DG) [3]. The tool adopts a generate-and-fix approach: first an instance is generated from the input schema, then it is validated against the schema using an external *validator* and, in case of validation failure, an attempt to fixing the instance is performed by exploiting the validation error messages. The generation for typed-assertions is straightforward but ignores interactions. This is related to a stronger limitation of not dealing with Boolean expressions: negation is simply ignored while binary operators, i.e. `allOf`, `anyOf` and `oneOf`, are handled by randomly selecting one branch while ignoring the other ones.

JSON Faker (JF) [2]. Differently from the previous one, this tool is not endowed with a fixing mechanism. Nevertheless, it has comparable limitations to the previous one and has even a very limited mechanism to deal with negated schemas and to merge typed-assertions of base values. On the down side, it adopts a random strategy for object generation which leads to generating objects with missing properties in some situations.

JSON Everything (JE) [1]. The last tool we select is part of a library for querying and managing JSONdata written in C#. It is based on a data generator for C# classes and is thus, limited in terms of the expressivity of the fragment of JSON Schema that is supported because object-oriented design does not support union type nor negation.

3 Preliminaries

3.1 The JSONData Model

In JSON, values can be of several types: `null`, `boolean`, `number`, `string`, `array` or `object`. Arrays are sorted lists of JSON values, the values can be of any type. Objects are key-value pairs, the keys l_i must be unique and of type `string`, values on the other hand can be any JSON value.

$$s \in \mathsf{Str}, d \in \mathsf{Num}, n \in \mathsf{Int}, n \geq 0, l_i \in \mathsf{Str}$$
$$J ::= \mathtt{null} \mid \mathtt{true} \mid \mathtt{false} \mid d \mid s \mid [\,J_1, \ldots, J_n\,] \mid \{\,l_1 : J_1, \ldots, l_n : J_n\,\} \; i \neq j \Rightarrow l_i \neq l_j$$

Example 1. Consider the following JSON object, consisting of four key-value pairs. The value associated with the key *"name"* is of type `string`. For the *"age"* key, it holds an `integer` value. The value of the key *"address"* is an object composed of 3 key-value pairs, all of which are of `string` type. Lastly, the value of *"hobbies"* is an `array` containing strings.

```
{
 "name": "John Doe",
 "age": 24,
 "address": {
         "street": "4 Pl. Jussieu",
         "city": "Paris",
         "zipCode": "75005"
 },
 "hobbies": ["Hiking","Cooking","Painting"]
}
```

3.2 Overview of JSON Schema

Many versions of JSON Schema have been released, the latest being Draft 2020-12, but the most widely adopted versions are anterior to that. A JSON Schema specification is itself a JSON document describing *typed-assertions* for enforcing specific kinds of JSONvalues (e.g objects, arrays, ...). Each typed-assertion uses the *type* keyword to indicate the kind of JSONvalue they relate to, and we will present later on their formal semantics by means of a set of validation rules. Before dealing with formal treatment, we give below an overall overview of assertions:

- Base values assertions are meant to describe strings or numbers: strings can be described by means of regular expressions using `pattern`, and their length can be restricted using an interval of integers (`minLength, maxLength`), while numbers can be captured by an interval of values (`minimum, maximum`) and by setting a number they should be multiple of (`multipleOf`).
- Array assertions are used for specifying the types of the elements that can appear in an array but also for restricting the size of the array itself using an interval (`minItems, maxItems`). It is also possible to require that a specific number of elements must satisfy a schema using the `contains, minContains,` and `maxContains` operators, and, finally, to disallow repetition of the same items using `uniqueItems`.
- Object assertions are used for specifying the schema of properties designated by their names or by regular expressions capturing their names (`properties, patternProperties`), for requiring the presence of some properties (`required`), for restricting the names used by properties (`propertyNames`), for limiting the number of properties using `minProperties` and `maxProperties`, and for expressing complex dependencies among properties (`dependentRequired` and `dependentSchemas`).

Besides these capabilities, the language allows for building complex expressions using Boolean operators like disjunction (`anyOf`), conjunction (`allOf`), and even negation (`not`). The language has also a variable definition mechanism using the `definitions` and `$ref` keywords which makes it convenient to reuse schema fragments within the same or an external schema. As a side effect, it is possible to specify a schema with recursive variables.

Example 2. To give an overview of assertions that can be expressed in JSON Schema, consider the following schema which restricts valid values to objects which must expose a property a whose value must be a string satisfying the pattern a(c|e)* and which admits additional properties whose values are not restricted in any way.

```
{ "type": "object",
  "properties": {"a": {"type": "string", "pattern": "a(c|e)*"}},
  "additionalProperties": true,
  "required": ["a"] }
```

In general, schemas tend to be more involved than the previous one and may feature some subtle interactions even between operators of the same assertion.

Example 3. The following schema extends the previous one with `patternProperties`. It enforces the values of the properties that are in the language of the regular expression `a+` to satisfy the assertion `minLength : 3`, and introduces, thus, an overlap with the pre-existing `properties` such that, a property `a` must satisfy both constraints, that is, be a string of at least three characters adhering to the pattern `a(c|e)*`.

```
{ "type": "object",
  "properties": {"a": {"type": "string", "pattern": "a(c|e)*"}},
  "required": ["a"],
  "patternProperties": {"a+": {"minLength": 3}} }
```

The above example, while being simple, is very helpful in characterizing the limitations of existing implementations of optimistic solutions, which, are unable to deal with overlapping constraints due to their limited support of *conjunction*.

Overlapping of constraints can also occur in array assertions like in the following example.

Example 4. The following specification describes arrays of at least four items, each consisting of a string of at least 2 characters adhering to the pattern `a(b|c)a`.

```
{ "type": "array",
  "items": {"type": "string" , "minLength": 2},
  "contains": {"pattern": "a(b|c)a"},
  "minItems": 4, "maxItems":10 }
```

3.3 Syntax and Semantics of JSON Schema

We formalize in Fig. 1 the syntax following Draft 2019 [14] by excluding keywords which introduce extra complexity to the validation which are `$anchor`, `$dynamicAnchor`, `$dynamicRef` and keywords whose semantics is based on annotations: unevaluatedProperties, unevaluatedItems.

We define the semantics of JSON Schema in an operational manner by considering the validation of an instance J over a schema S which yields a boolean. For convenience, we define this validation using a set of rules which are guided by the syntax of JSON Schema and which rely on a main judgment and two auxiliary judgments, all of which are mutually dependent.

- the *schema judgment* is the main one, denoted $J ~?~ S \mapsto r$, captures the validation of J against S and returns the boolean value r;
- the auxiliary judgements, *keyword judgment* denoted $J ~?~ K \to r$, and *keywords-list judgment* denoted $J ~?~ \boldsymbol{K} \Rightarrow r$, are meant to capture the validation of J over a single keyword K or a list of keywords $\boldsymbol{K}$, respectively.

$q \in \mathsf{Num}, i \in \mathsf{Int}, k \in \mathsf{Str}, uri \in \mathsf{Str}, for \in \mathsf{Str}, p \in \mathsf{Str}, J \in \mathit{JVal}$
$Tp ::= \texttt{object} \mid \texttt{number} \mid \texttt{integer} \mid \texttt{string} \mid \texttt{array} \mid \texttt{boolean} \mid \texttt{null}$
$S \ ::= \texttt{true} \mid \texttt{false} \mid \{\ Kw\ (,Kw)^*\ \} \mid \{\ \}$
$Kw ::= \texttt{type}: Tp \mid \texttt{type}: [Tp_1, \ldots, Tp_n] \mid \texttt{minimum}: q \mid \texttt{maximum}: q$
$\quad \mid \texttt{exclusiveMinimum}: q \mid \texttt{exclusiveMaximum}: q \mid \texttt{multipleOf}: q$
$\quad \mid \texttt{pattern}: p \mid \texttt{minLength}: i \mid \texttt{maxLength}: i \mid \texttt{format}: for$
$\quad \mid \texttt{minProperties}: i \mid \texttt{maxProperties}: i \mid \texttt{required}: [k_1, \ldots, k_n]$
$\quad \mid \texttt{properties}: \{\ k_1: S_1, \ldots, k_m: S_m\ \} \mid \texttt{additionalProperties}: S$
$\quad \mid \texttt{patternProperties}: \{\ p_1: S_1, \ldots, p_m: S_m\ \} \mid \texttt{propertyNames}: S$
$\quad \mid \texttt{dependentSchemas}: \{\ k_1: S_1, \ldots, k_n: S_n\ \}$
$\quad \mid \texttt{dependentRequired}: \{\ k_1: [k_1^1, \ldots, k_{o_1}^1], \ldots, k_m: [k_1^n, \ldots, k_{o_n}^n]\ \}$
$\quad \mid \texttt{minItems}: i \mid \texttt{maxItems}: i \mid \texttt{minContains}: i \mid \texttt{maxContains}: i$
$\quad \mid \texttt{uniqueItems}: b \mid \texttt{items}: S \mid \texttt{prefixItems}: [S_1, \ldots, S_n] \mid \texttt{contains}: S$
$\quad \mid \texttt{\$defs}: \{\ k_1: S_1, \ldots, k_n: S_n\ \} \mid \texttt{\$ref}: uri \mid \texttt{anyOf}: [S_1, \ldots, S_n]$
$\quad \mid \texttt{allOf}: [S_1, \ldots, S_n] \mid \texttt{oneOf}: [S_1, \ldots, S_n] \mid \texttt{not}: S$
$\quad \mid \texttt{if}: S \mid \texttt{else}: S \mid \texttt{then}: S \mid \texttt{const}: J_c \mid \texttt{enum}: [J_1, \ldots, J_n]$

Fig. 1. Grammar of normalized JSON Schema Draft 2019.

This latter judgement is useful for defining the semantics of keywords whose semantics depends on the evaluation of adjacent keywords like in the case of `additionalProperties`.

We follow the terminology of JSON Schemastandard by distinguishing between terminal assertions, boolean applicators and structural applicators.

3.3.1 Terminal Assertions

These assertions assemble keywords that do not contain any subschema. First, we have the *type-uniform* ones that do not make a differentiation based on the type of the instances, these are `enum`, `const`, $\texttt{type}: [\mathsf{Tp}_1, \ldots, \mathsf{Tp}_n]$, and $\texttt{type}: \mathsf{Tp}$. Then, we have the implicative keywords, the ones that are specific to a certain type T which always return $\mathcal{T}$ (true) when applied to instances whose types differ from T. To capture the semantics of a terminal assertion $kw: J'$ when applied to an instance J, we consider two validation rules ($kw\mathsf{Triv}$) and (kw), the former rule is for the trivial case of the typed assertions, and the latter is for both typed assertions in the general case and the type-uniform assertions, for the type-uniform ones the condition $TypeOf(J) = TypeOf(kw)$ is ignored, where $TypeOf(J)$ extracts the type of J, while $TypeOf(kw)$ indicates the type to which kw refers. Table 1 presents all the terminal assertions and specifies the condition verified for each assertion when applied to the instance J.

$$\frac{TypeOf(J) \neq TypeOf(kw)}{J\ ?\ kw: J' \rightarrow \mathcal{T}} \qquad (kw\mathsf{Triv})$$

$$\frac{TypeOf(J) = TypeOf(kw) \quad r = \mathsf{cond}(J, kw:J')}{J\ ?\ kw:J' \to r} \qquad (kw)$$

Table 1. Terminal assertions: conditions

assertion kw:J'	$TypeOf(kw)$	cond(J,kw:J')
enum : $[J_1, \ldots, J_n]$	no type	$J \in \{\!\!\{J_1, \ldots, J_n\}\!\!\}$
const : J_c	no type	$J = J_c$
type : Tp	no type	$TypeOf(J) = $ Tp
type : $[\mathsf{Tp}_1, \ldots, \mathsf{Tp}_n]$	no type	$TypeOf(J) \in \{\!\!\{\mathsf{Tp}_1, \ldots, \mathsf{Tp}_n\}\!\!\}$
exclusiveMinimum: q	number	$J > q$
exclusiveMaximum: q	number	$J < q$
minimum: q	number	$J \geq q$
maximum: q	number	$J \leq q$
multipleOf: q	number	$\exists i \in $ Int. $J = i \times q$
pattern: p	string	$J \in L(p)$
minLength: i	string	$|J| \geq i$
maxLength: i	string	$|J| \leq i$
minProperties: i	object	$|J| \geq i$
maxProperties: i	object	$|J| \leq i$
required : $[k_1, \ldots, k_n]$	object	$\forall i.\ k_i \in \mathtt{names}(J)$
uniqueItems: true	array	$J = [J_1, \ldots, J_n]$ with $n \geq 0$ $\wedge \forall i,j.\ 1 \leq i \neq j \leq n \Rightarrow J_i \neq J_j$
uniqueItems: false	array	T
minItems: i	array	$|J| \geq i$
maxItems: i	array	$|J| \leq i$
format: $format$	string	$J \in L(format)$
dependentRequired : $\{\ k_1 : [k_1^1, \ldots, k_{o_1}^1],$ $\ldots, k_n : [k_1^n, \ldots, k_{o_n}^n]\ \}$	object	$\forall i \in \{\!\!\{1 \ldots n\}\!\!\}.$ $\quad k_i \in \mathtt{names}(J)$ $\Rightarrow \{\!\!\{k_1^i, \ldots, k_{o_i}^i\}\!\!\} \subseteq \mathtt{names}(J)$

3.3.2 Boolean Applicators

The semantics of the boolean operators is straightforward and relies on applying the corresponding logical connectives over the validation result of each schema.

$$\frac{J\ ?\ S \mapsto r}{J\ ?\ \mathtt{not}:S \to \neg r} \qquad (\mathrm{not})$$

$$\frac{\forall i \in \{\!\!\{1 \ldots n\}\!\!\}.\ J\ ?\ S_i \mapsto r_i \quad r = \vee(\{\!\!\{r_i\}\!\!\}^{i \in \{\!\!\{1 \ldots n\}\!\!\}})}{J\ ?\ \mathtt{anyOf}:[S_1, \ldots, S_n] \to r} \qquad (\mathrm{anyOf})$$

$$\frac{\forall i \in \{\!|1\ldots n|\!\}.\ J\ ?\ S_i \mapsto r_i \qquad r = \wedge(\{\!|r_i|\!\}^{i\in \{\!|1\ldots n|\!\}})}{J\ ?\ \texttt{allOf} : [S_1, ..., S_n] \to r} \quad \text{(allOf)}$$

$$\frac{\forall i \in \{\!|1\ldots n|\!\}.\ J\ ?\ S_i \mapsto r_i \qquad r = (\ |\{\!|i \mid r_i = T|\!\}| = 1\ )}{J\ ?\ \texttt{oneOf} : [S_1, ..., S_n] \to r} \quad \text{(oneOf)}$$

3.3.3 Structural Operators

We distinguish between three families of operators, those pertaining to *objects*, *arrays* and the rest. For each family, we further distinguish between independent operators whose semantics is described using the *keyword* judgment and dependent operators whose semantics depends on the evaluation of adjacent operators and are hence captured using the *keywords-list* judgment.

Object Keywords. These keywords are specific to the object type, hence, their semantics are trivial and they are always satisfied when applied to instances that are not of type object. They are composed of subschemas, thus, they require the verification of the internal structures of the instance.

First, to illustrate the trivial behavior of these keywords when encountering an instance that is not of type object, we present the trivial rule for the `properties` keyword, that is:

$$\frac{TypeOf(J) \neq \textsf{object}}{J\ ?\ \texttt{properties} : \{\ k_1 : S_1, \ldots, k_m : S_m\ \} \to T} \quad \text{(propertiesTriv)}$$

The (properties) rule captures the non-trivial case, that is, when J is an object of the form $J = \{\ k'_1 : J_1, \ldots, k'_n : J_n\ \}$. This rule states that, in order for J to validate $\texttt{properties} : \{\ k_1 : S_1, \ldots, k_m : S_m\ \}$, every value J_i whose key k'_i matches a keyword $k_j : S_j$, must validate the schema S_j. This rule relies on collecting the set of indexes of matching pairs in order to perform validation whose result is combined using conjunction.

$$\frac{\begin{array}{c}J = \{\ k'_1 : J_1, \ldots, k'_n : J_n\ \} \qquad \{\!|(i_1, j_1), \ldots, (i_l, j_l)|\!\} = \{\!|(i,j) \mid k'_i = k_j|\!\} \\ \forall q \in \{\!|1\ldots l|\!\}.\ J_{i_q}\ ?\ S_{j_q} \mapsto r_q \qquad r = \wedge(\{\!|r_q|\!\}^{q\in\{\!|1\ldots l|\!\}})\end{array}}{J\ ?\ \texttt{properties} : \{\ k_1 : S_1, \ldots, k_m : S_m\ \} \to r} \quad \text{(properties)}$$

The (patternProperties) rule generalizes the previous one by considering key-pattern *membership* expressed by $k'_i \in L(p_j)$ where $L(p_j)$ denotes the language of the pattern p_j, instead of *exact* label-keyword matching. The same conjunctive semantics used in the previous rule is adopted here.

$$\frac{\begin{array}{c}J = \{\ k'_1 : J_1, \ldots, k'_n : J_n\ \} \qquad \{\!|(i_1, j_1), \ldots, (i_l, j_l)|\!\} = \{\!|(i,j) \mid k'_i \in L(p_j)|\!\} \\ \forall q \in \{\!|1\ldots l|\!\}.\ J_{i_q}\ ?\ S_{j_q} \mapsto r_q \qquad r = \wedge(\{\!|r_q|\!\}^{q\in\{\!|1\ldots l|\!\}})\end{array}}{J\ ?\ \texttt{patternProperties} : \{\ p_1 : S_1, \ldots, p_m : S_m\ \} \to r} \quad \text{(patternProperties)}$$

The (propertyNames) rule captures the semantics of `propertyNames` which, differently from other operators which constrain the instance values based on a specific schema, constrains the instance labels by imposing them to adhere to the associated schema S. The semantics of `propertyNames` is trivially conjunctive, as expressed in the premise of the rule.

$$\frac{J = \{\!|\, k_1 : J_1, \ldots, k_n : J_n \,|\!\} \quad \forall i \in \{\!|1\ldots n|\!\}.\ k_i\ ?\ S \mapsto r_i \quad r = \wedge(\{\!|r(\sigma_i)|\!\}^{i \in \{\!|1\ldots n|\!\}})}{J\ ?\ \mathtt{propertyNames} : S \to r} \quad \text{(propertyNames)}$$

The semantics of `dependentSchemas` is conditional and allows applying a sub-schema in case a label exists. The corresponding rule expresses this fact by collecting all sub-schemas that need to be applied on the validated instance J and combining the validation results with a conjunction.

$$\frac{J = \{\!|\, k'_1 : J_1, \ldots, k'_m : J_m \,|\!\} \quad \{\!|i_1, \ldots, i_l|\!\} = \{\!|i \mid i \in \{\!|1\ldots n|\!\},\ k_i \in \{\!|k'_1, \ldots, k'_m|\!\}|\!\}}{\forall q \in \{\!|1\ldots l|\!\}.\ J\ ?\ S_{i_q} \mapsto r_q \quad r = \wedge(\{\!|r_q|\!\}^{q \in \{\!|1\ldots l|\!\}})} \quad \text{(dependentSchemas)}$$
$$J\ ?\ \mathtt{dependentSchemas} : \{\, k_1 : S_1, \ldots, k_n : S_n \,\} \to r$$

The `additionalProperties` operator is evaluated relatively to a context where any value J_{i_q} which has not been validated by an adjacent `properties` or `patternProperties` must adhere to the schema S. The evaluated values are those remaining after eliminating the pairs whose keys are in the languages of the properties and patterns extracted by the function `propsOf(K)`. The function `propsOf` is defined as follows, where the notation $\underline{k_i}$ indicates a pattern whose language accepts only the string value k_i:

$$\begin{array}{ll} \mathtt{propsOf}(\mathtt{properties} : \{\, k_1 : S_1, \ldots, k_m : S_m \,\}) & = \underline{k_1} \cdot | \cdot \ldots \cdot | \cdot \underline{k_n} \\ \mathtt{propsOf}(\mathtt{patternProperties} : \{\, p_1 : S_1, \ldots, p_m : S_m \,\}) & = p_1 \cdot | \cdot \ldots \cdot | \cdot p_m \\ \mathtt{propsOf}(K) & = \emptyset \qquad \text{otherwise} \\ \mathtt{propsOf}([\, K_1, \ldots, K_n \,]) & = \mathtt{propsOf}(K_1) \cdot | \cdot \ldots \cdot | \cdot \mathtt{propsOf}(K_n) \end{array}$$

The semantics of this operator is also conjunctive, and the associated rule is a specific case of the (KW-List) rule which collects the validation results of a keyword-list $\boldsymbol{K}$ in a conjunctive manner.

$$\frac{\begin{array}{c} J = \{\!|\, k_1 : J_1, \ldots, k_n : J_n \,|\!\} \qquad J\ ?\ \boldsymbol{K} \Rightarrow r \\ \{\!|i_1, \ldots, i_l|\!\} = \{\!|i \mid 1 \leq i \leq n \wedge k_i \notin L(\mathtt{propsOf}(\boldsymbol{K}))|\!\} \\ \forall q \in \{\!|1\ldots l|\!\}.\ J_{i_q}\ ?\ S \to r_q \quad r' = \wedge(\{\!|r_q|\!\}^{q \in \{\!|1\ldots l|\!\}}) \end{array}}{J\ ?\ (\boldsymbol{K} + \mathtt{additionalProperties} : S) \Rightarrow r \wedge r'} \quad \text{(additionalProperties)}$$

Array Keywords. These keywords only impact instances of type array, otherwise, their effect is trivial. Similar to the object keywords, evaluating an instance against such keywords necessitates the verification of the instance's internal structure.

The following rule evaluates array-type instances against a set of schemas specified within the prefixItems keyword. It asserts that each item of the instance located at position i, where $i \leq n$, is validated against the ith schema of prefixItems. The overall validity of the instance is obtained by combining the boolean values returned for each item. When either the instance or the list of schemas is empty, the returned value is $\mathcal{T}$.

$$\frac{J = [J_1, \ldots, J_m] \quad \forall i \in \{\!|1 \ldots \min(n,m)|\!\}.\ J_i\ ?\ S_i \mapsto r_i \quad r = \wedge(\{\!|r_i|\!\}^{i \in \{\!|1 \ldots \min(n,m)|\!\}})}{J\ ?\ \text{prefixItems} : [S_1, \ldots, S_n] \to r}$$

(prefixItems)

The (contains) rule checks whether an item of the instance satisfies the schema S. Consequently, it returns $\mathcal{T}$ if at least one item fulfills the condition, and $\mathcal{F}$ otherwise.

$$\frac{J = [J_1, \ldots, J_n] \quad \forall i \in \{\!|1 \ldots n|\!\}.\ J_i\ ?\ S \mapsto r_i \quad r = \vee(\{\!|r_i|\!\}^{i \in \{\!|1 \ldots n|\!\}})}{J\ ?\ \text{contains} : S \to r}$$

(contains)

The following rule checks the validity of the items that have not been evaluated by an adjacent prefixItems keyword present in $\boldsymbol{K}$. The list of items to verify are all those whose indexes come after the index captured by the function $maxPrefixOf$, which is defined as follows:

```
maxPrefixOf(prefixItems : [S_1, ..., S_m]) = m
maxPrefixOf(K)                            = 0                    otherwise
maxPrefixOf([K_1, ..., K_n])              = max_{i∈{|1...n|}} maxPrefixOf(K_i)
```

The overall validity of the instance is obtained by combining the boolean values returned for each item.

$$\frac{\begin{array}{c} J = [J_1, \ldots, J_n] \quad J\ ?\ \boldsymbol{K} \Rightarrow r \quad \{\!|i_1, \ldots, i_l|\!\} = \{\!|1 \ldots n|\!\} \setminus \{\!|1 \ldots maxPrefixOf(\boldsymbol{K})|\!\} \\ \forall q \in \{\!|1 \ldots l|\!\}.\ J_{i_q}\ ?\ S \mapsto r_q \quad r' = \wedge(\{\!|r_q|\!\}^{q \in \{\!|1 \ldots l|\!\}}) \end{array}}{J\ ?\ (\boldsymbol{K} + \text{items} : S) \Rightarrow r \wedge r'}$$

(items)

The (contains–bounds) rule captures the semantics of contains in presence of the minContains and maxContains keywords. It generalizes the contains rules by introducing cardinality constraints: an array is valid when the cardinality of the items satisfying contains ranges between the values of minContains and maxContains.

$$\frac{J = [J_1, \ldots, J_n] \quad \forall i \in \{\!|1 \ldots n|\!\}.\ J_i\ ?\ S \mapsto r_i \quad \kappa_c = \{\!|i \mid r_i = \mathcal{T}|\!\} \quad r_c = (i \leq |\kappa_c| \leq j)}{J\ ?\ (\text{contains} : S + \text{minContains} : i + \text{maxContains} : j) \Rightarrow r_c}$$

(contains–bounds)

Other Keywords. The $ref rule states the obvious validation of a $ref: uri statement which amounts to validating schema S' referenced by uri.

The two remaining rules capture the validation of conditionals which relies on validating either the *then* or *else* sub-schema depending on the validation result of the *if* statement.

$$\frac{S' = \texttt{getSchema}(uri) \qquad J \;?\; S' \mapsto r}{J \;?\; \texttt{\$ref} : uri \to r} \quad (\$\text{ref})$$

$$\frac{J \;?\; K \Rightarrow r \qquad J \;?\; S_i \mapsto \mathcal{T} \qquad J \;?\; S_t \mapsto r'}{J \;?\; (K + \texttt{if} : S_i + \texttt{then} : S_t + \texttt{else} : S_e\;) \Rightarrow r \wedge r'} \quad (\text{if} - \text{true} - \text{then})$$

$$\frac{J \;?\; K \Rightarrow r \qquad J \;?\; S_i \mapsto \mathcal{F} \qquad J \;?\; S_e \mapsto r'}{J \;?\; (K + \texttt{if} : S_i + \texttt{then} : S_t + \texttt{else} : S_e\;) \Rightarrow r \wedge r'} \quad (\text{if} - \text{false} - \text{else})$$

4 The Pre-processing Phase

Our witness generation follows the spirit of optimistic solutions by avoiding the complex schema rewritings and preparation mechanisms developed in [8]. Unlike other optimistic solutions that may use non-deterministic choices, our solution ensures soundness by sacrificing completeness: it returns a valid witness if the schema is satisfiable and indicates schema *emptiness* if the schema is not satisfiable, but may fail at processing a schema if it is too complex or uses conjunction/negation that we cannot eliminate.

To achieve soundness, our solution manages conjunction and negation consistently by following the *canonicalization* mechanism developed in [11]. It relies on a pre-processing phase meant to *normalize* the input schema and facilitate the *generation* phase. This phase expands references up to a certain depth and normalizes the expanded schema, producing an equivalent one in a canonical form that is more amenable to witness generation. Schema expansion is straightforward, while schema normalization follows the rules of Habib et al. [11], which we adapt to support Draft 2019 of the standard [14].

4.1 Reference Expansion

This consists in recursively substituting, a finite number of times, every occurence of a reference with the fragment that it refers to.

Example 5. To illustrate how references are expanded, consider a recursive schema describing persons by means of their names and their potentially non empty list of children which is represented by an array of persons so defined.

```
{ "type": "object",
  "properties": {"name": { "type": "string" },
   "children": {"type": "array","items": { "$ref": "#" } }
  }
}
```

The expansion, up to one level of depth, produces a schema where the nested "children" property has an empty array so as to indicate termination on the recursive branch expressed inside the items assertion.

```
{ "type": "object",
  "properties": {"name": { "type": "string" },
  "children": {"type": "array",
    "items":{"type": "object",
        "properties": {"name": {"type": "string" },
        "children": {"type": "array","items": false}} }
          } }
}
```

This expansion is described below, by focusing on the main cases, where S_b denotes the *base schema* that is called initially and which serves for resolving (local) references appearing at an arbitrary nesting level. We also use a context C to register the references that have already been expanded to detect any cycle.

$Expan(\{\texttt{\$ref}:f\},C) ::= Expan(gets(S_b,f),C\cup\{f\})$ if $f\notin C$
$Expan(\{\texttt{\$ref}:f\},C) ::= \texttt{false}$ if $f\in C$
$Expan(\{\texttt{not}:S\},C) ::= \{\texttt{not}:Expan(S,C)\}$
$Expan(\{\texttt{anyOf/allOf/oneOf}:[S_1,..,S_n]\},C) ::=$
$\quad\{\texttt{anyOf/allOf/oneOf}:[Expan(S_1,C),..,Expan(S_n,C)]\}$

4.2 Schema Normalization

Schema normalization is meant to rewrite the original schema into an equivalent schema which is more convenient to use for generation. It consists in applying the following semantics-preserving transformations:

– minimize *syntactic sugar* by removing redundant operators in object and array assertions and rewriting `oneOf` into its logical equivalent;
– group expressions pertaining to the same type into a single *typed assertion* and also perform standard simplifications. This involves eliminating inconsistent branches, like those involving *type* assertions with distinct types or incompatible bounds for number assertions;
– eliminate negation in the two following cases: through *string* assertions, in the general case, and through *number* assertions, when `multipleOf` is absent. In other situations, complete negation elimination requires to extend the schema language with additional operators as described in [9].
– merge *similar-type* assertions, by combining their constraints and detect *emptiness*, in specific cases.

$$\frac{S_1 = \{\,\texttt{type}: \mathsf{T}, K\,\} \quad S_2 = \{\,\texttt{type}: \mathsf{T'}, K'\,\} \quad \mathsf{T} \neq \mathsf{T'}}{\{\,\texttt{allOf}: [\,S_1, S_2\,]\,\} \to \mathit{false}} \text{ (heterogenous types)}$$

$$\frac{\begin{array}{c} S_i = \{\,\texttt{type}: \texttt{number}, \texttt{minimum}: m_i, \texttt{maximum}: M_i, \texttt{multipleOf}: \mathit{mof}_i\,\} \\ i = 1, 2 \quad m = \max(m_1, m_2) \quad M = \min(M_1, M_2) \quad l = \mathit{lcm}(\mathit{mof}_1, \mathit{mof}_2) \end{array}}{\{\,\texttt{allOf}: [\,S_1, S_2\,]\,\} \to \{\,\texttt{type}: \texttt{number}, \texttt{minimum}: m, \texttt{maximum}: M, \texttt{multipleOf}: l\,\}}$$
(intersect number)

$$\frac{S_i = \{\,\texttt{type}: \texttt{string}, \texttt{pattern}: p_i\,\} \quad i = 1, 2}{\{\,\texttt{allOf}: [\,S_1, S_2\,]\,\} \to \{\,\texttt{type}: \texttt{string}, \texttt{pattern}: p_1 \cap p_2\,\}} \text{ (intersect string)}$$

$$\frac{\begin{array}{c} S_i = \{\texttt{type}: \texttt{array}, \texttt{minItems}: m_i, \texttt{maxItems}: M_i, \texttt{prefixItems}: [S_1^i, \ldots S_{n_i}^i], \texttt{items}: S_i', \\ \texttt{contains}: S_{c_i}, \texttt{minContains}: mc_i, \texttt{maxContains}: Mc_i\} \\ i = 1, 2 \quad m = \max(m_1, m_2) \quad M = \min(M_1, M_2) \\ \mathit{pItems} = \mathit{mergeItems}([S_1^1, \ldots S_{n_1}^1], [S_1^2, \ldots S_{n_2}^2]) \end{array}}{\begin{array}{c} \{\,\texttt{allOf}: [\,S_1, S_2\,]\,\} \to \{\texttt{type}: \texttt{array}, \texttt{minItems}: m, \texttt{maxItems}: M, \\ \texttt{prefixItems}: \mathit{pItems}, \texttt{items}: \{\,\texttt{allOf}: [\,S_1', S_2'\,]\,\}, \\ \texttt{allOf}: [\{\,\texttt{contains}: S_{c_1}, \texttt{minContains}: mc_1, \texttt{maxContains}: Mc_1\,\}, \\ \{\,\texttt{contains}: S_{c_2}, \texttt{minContains}: mc_2, \texttt{maxContains}: Mc_2\,\}]\} \end{array}}$$
(intersect array)

$$\frac{\begin{array}{c} S_i = \{\texttt{type}: \texttt{object}, \texttt{minProperties}: m_i, \texttt{maxProperties}: M_i, \texttt{required}: [k_1^i, \ldots, k_n^i], \\ \texttt{patternProperties}: \{p_1^i: S_1^i, \ldots, p_{l_i}^i: S_{l_i}^i\}, \texttt{propertyNames}: S^i\} \\ i = 1, 2 \quad m = \max(m_1, m_2) \quad M = \min(M_1, M_2) \\ \mathit{pattProps} = \mathit{mergeProps}(\{p_1^1: S_1^1, \ldots, p_{l_1}^1: S_{l_1}^1\}, \{p_1^2: S_1^2, \ldots, p_{l_2}^2: S_{l_2}^2\}) \end{array}}{\begin{array}{c} \{\,\texttt{allOf}: [\,S_1, S_2\,]\,\} \to \{\texttt{type}: \texttt{object}, \texttt{minProperties}: m, \texttt{maxProperties}: M, \\ \texttt{required}: [k_1, \ldots, k_n \cup] \cup [k_1', \ldots, k_l'], \\ \texttt{patternProperties}: \mathit{pattProps}, \texttt{propertyNames}: \{\,\texttt{allOf}: [\,S, S'\,]\,\}\} \end{array}}$$
(intersect object)

Fig. 2. Conjunction elimination rules

Schema normalization is a merely technical phase of our approach which relies on rules which follow the lines of those defined in [11]. We provide only the *merge* rules which are useful for our presentation and part of which we adapted to conform with the draft of JSON Schemawhich is more recent than the one used in [11].

These merge rules are presented in Fig. 2. The first rule is trivial, and captures the merge of schemas referring to incompatible types. Each of the remaining rules is dedicated to a specific type and is based on combining the constraints of this type while adhering to it semantics. For instance, for number types, the bounds are intersected and the `multipleOf`arguments combined using lcm while for string types, the patterns, if any, are intersected. The array rule uses a similar logic for bounds, and relies on combining the `prefixItems` sub-schemas by calling $\mathit{mergeItems}$ which takes the longest common prefix, and resorts to `items` for the sub-schemas in the positions outside the common prefix. The `contains` constraints, which are existential by nature, are combined into `allOf` to preserve their semantics. The logic for object types is rather obvious and relies on intersecting the bounds, collecting the required labels of both schemas, and combining the properties by merging the schemas associated to the same patterns using $mergeProps$.

The normalized schemas are in a disjunctive normal form whose conjuncts may be negated due to the impossibility to eliminate negation of object and array assertions and of the `multipleOf` operator. The general form of the normalized schemas is captured by the grammar of Fig. 3 which describes the general schemas S in DNF, and the form of the conjuncts $\mathcal{S}$ which are built starting from the positive *typed assertions* TA or the negative ones, denoted with $notTA$ and which concern object and array assertions only. For number assertions, negation can still appear in front of `multipleOf` as expected.

5 The Generation Phase

We recall that our goal is to generalize the *witness generation* problem to that of generating N *valid* instances adhering to a schema S. Our optimistic solution is characterized by its *incompleteness* since, for some class of schemas, it is not able to process the input schema due to a request for processing a negated object or array sub-schema. When our solution succeeds in processing the input schema, it yields a (possibly empty) set of $M \leq N$ *valid* instances.

The generation proceeds by case analysis on a schema adhering to the *generation grammar* and is captured by two mutually recursive functions Gen, dealing with schemas in the partially disjunctive normal form denoted with S, and $GenS$,

$q \in \mathsf{Num}, i \in \mathsf{Int}, k \in \mathsf{Str}, p \in \mathsf{Str}, J \in \mathsf{JVal}$

$S \quad ::= \mathtt{true} \mid \mathtt{false} \mid \mathcal{S} \mid \{\, \mathtt{anyOf} : [\, \mathcal{S}\,(,\mathcal{S})^+\,]\,\}$

$\mathcal{S} \quad ::= \{\,\mathtt{allOf} : [\, TA^? \,(, NotTA)^+\,]\,\} \mid \{\, TA \,(, \mathtt{enum} : J)^?\,\} \mid notTA$

$TA \quad ::= NullTA \mid BoolTA \mid NumTA \mid IntTA \mid StrTA \mid ArrTA \mid ObjTA$

$NullTA ::= \mathtt{type} : \mathtt{null}$

$BoolTA ::= \mathtt{type} : \mathtt{boolean}$

$NumTA ::= \mathtt{type} : \mathtt{number}\,(,\mathtt{minimum} : q)^?(,\mathtt{maximum} : q)^?(,\mathtt{multipleOf} : q)^?$
$\qquad\qquad (,\mathtt{not} : \{\,\mathtt{multipleOf} : q\,\})^?$

$IntTA \quad ::= -\mathtt{type} : \mathtt{integer}\,(, TermTA)^?$

$StrTA \quad ::= \mathtt{type} : \mathtt{string}\,(, \mathtt{pattern} : p)^?$

$ArrTA \quad ::= \mathtt{type} : \mathtt{array}\,(, \mathtt{minItems} : i)^?(, \mathtt{maxItems} : i)^?$
$\qquad\qquad (, \mathtt{uniqueItems} : b)^?(, \mathtt{items} : S)^?(, \mathtt{prefixItems} : [\, S_1, \ldots, S_n\,])^?$
$\qquad\qquad (contS \mid \{\,\mathtt{allOf} : [\, contS(, contS)^+\,]\,\})$

$ObjTA \quad ::= \mathtt{type} : \mathtt{object}\,(, \mathtt{minProperties} : i)^?(, \mathtt{maxProperties} : i)^?(, \mathtt{required} : [\,k_1, \ldots, k_n\,])^?$
$\qquad\qquad (, \mathtt{patternProperties} : \{\, p_1 : S_1, \ldots, p_m : S_m\,\})^?(, \mathtt{propertyNames} : S)^?$

$notTA \quad ::= \{\,\mathtt{not} : ObjTA\,\} \mid \{\,\mathtt{not} : ArrTA\,\}$

$contS \quad ::= (, \mathtt{minContains} : i)^?(, \mathtt{maxContains} : i)^?(, \mathtt{contains} : S)^?$

Fig. 3. Generation grammar

dealing with the disjuncts denoted with $\mathcal{S}$. Both functions are defined in Fig. 4. The first line of Gen deals with the universally satisfied schema true and yields a default set of values, consisting in integers ranging from 1 to N. The second line deals with the unsatisfiable schema false whose generation yields, obviously, an empty set. The third line deals with the disjunctive expression and relies on generating values from one of its disjuncts, the one producing the largest set of values close to N. In case no assertion yields a result, generation simply fails.

The logic of $GenS$ is straightforward: generation fails in presence of negated assertions (first line and third line), uses validation in the presence of enum to select valid values only, while restricting to N arbitrarily chosen values using $trunc$, in case $n > N$ (second line), and resorts to calling specific algorithms for generating types assertions pertaining to the four types (last line).

5.1 Generation of Basic Types

– type *"null"*: only the $null$ value can be generated.
– type *"bool"*: we either generate $false$, $true$ or both values.
– type *"number"*: the generation for numbers generalizes the function defined in the witness generation approach [8]. In that work, generating a number given a number assertion produces a single valid value while we aim at generating N distinct numbers. To do so, we adopt a fairly straightforward iterative approach based on returning values within the minimum, maximum interval, satisfying the argument of multipleOf and violating the argument of a negated multipleOf. Each iteration updates the interval bounds by adding or subtracting an ϵ value determined a priori by analyzing the interval length and the cardinality N. Updating the values of the bounds is meant to exclude the previously generated value v_i. If the minimum constraint is present, its value is updated to $v_i + \epsilon$; otherwise, if maximum is the only bound constraint present in the schema, then its value is updated to $v_i - \epsilon$. If neither of them is present, the minimum constraint is added, and its value is instantiated to $v_i + \epsilon$.
– type *"string"*: the generation of N distinct words is delegated to the Brics library whose limitation forced us to adopt a technically involved approach: we instantiate an initial automaton $\mathcal{A}$ that accepts the language of the regular expression of the pattern constraint. Then, we iterate N times, and whenever we generate a new word w, we exclude it from the automaton $\mathcal{A}$ to prevent its

$Gen(\text{true}, N)$ $::= \{\!\{ 1, \ldots, N \}\!\}$
$Gen(\text{false}, N)$ $::= \emptyset$
$Gen(\{\, \text{anyOf} : [\mathcal{S}_1, \ldots, \mathcal{S}_n\,]\, \}, N)$ $::= GenS(\mathcal{S}_i, N)$ s.t. $GenS(\mathcal{S}_i, N) \neq fail$ and for $i, j = 1..n$
$\quad |GenS(\mathcal{S}_i, N)| = Min(N, Max_j\{\, |GenS(\mathcal{S}_j, N)|\, \})$
$\quad fail$ otherwise

$GenS(\{\, \text{allOf} : [\, TA, NotTA\,]\, \}, N)$ $::= fail$
$GenS(\{\, TA, \text{enum} : [\, J_1, \ldots, J_n\,]\, \}, N) ::= trunc(N, \{\!\{ J_i \mid J_i \,?\, TA \mapsto true \}\!\})$
$GenS(NotTA, N)$ $::= fail$
$GenS(TA, N)$ $::=$ see specific algorithms

Fig. 4. Instance generation: main algorithm

regeneration. This is achieved by updating the automaton $\mathcal{A}$ s.t. $\mathcal{A} = \mathcal{A} \cap \mathcal{A}'$, where $\mathcal{A}'$ is the complement of the automaton that accepts only w.

5.2 Object Types

Given an object schema

$ObjTA ::=$
$\{$ type : object, minProperties : min, maxProperties : max,
    patternProperties :$\{p_1 : S_1, ..., p_n : S_n\}$,
    required :$[l_1, ..., l_k]$
    propertyNames : S_{str} $\}$

and the desired number N of distinct objects, generation is performed by the following Algorithm 1. It first invokes SatReq(cf. Algorithm 2) in order to generate the values of the *required* fields. In case it is not possible to generate a value for at least one of the required fields, generation *fails*. Otherwise generation goes on by calling SatMinProp(cf. Algorithm 3) which checks whether minProperties is satisfied and, if not, it generates additional fields by attempting to produce enough values for each additional field so as to increase the number of possible objects that can be obtained. Then, if the sets of generated values for fields satisfying minProperties are sufficient to obtain at least N objects, we return the generated instances; Otherwise, if it is possible to add new fields or replace non-required existing ones (as verified by |required|< maxProperties), MoreInstancesfunction (cf. Algorithm 4) is called to generate the missing instances. This is achieved by adding more fields while adhering to the maxProperties constraint in order to reach N objects.

Algorithm 1: main object generation algorithm

Data: An object assertion $ObjTA$, an integer N
Result: $M \leq N$ instances | *fail*
1 $Res = \{\}$
2 $objMap = SatReq(ObjTA, N)$
3 **if** $objMap \neq fail$ **then**
4 $\quad$ $objMap = SatMinProp(ObjTA, objMap, N)$
5 $\quad$ **if** $objMap \neq fail$ **then**
6 $\quad\quad$ $Res = objProduct(objMap)$
7 $\quad\quad$ $M = |Res|$
8 $\quad\quad$ **if** $M < N$ *and* $|required| <$ maxProperties **then**
9 $\quad\quad\quad$ **return** $Res \cup MoreInstances(ObjTA, objMap, N-M)$
10 $\quad\quad$ **else**
11 $\quad\quad\quad$ **return** Res

12 **return** fail

Example 6. To illustrate the generation of objects, consider the following normalized object schema which requires every valid object to have at least 4 properties, two of which must have the keys "a" and "b".

```
{ "type": "object", "minProperties": 4, "maxProperties": 5,
  "patternProperties": {
          "a": {"type": "integer", "minimum": 1, "maximum": 4},
          "b": {"type": "integer", "minimum": 1, "maximum": 1},
          "c": {"type": "integer", "minimum": 1, "maximum": 5},
          "d": {"type": "integer", "minimum": 1, "maximum": 4},
          "e": {"type": "integer", "minimum": 1, "maximum": 4},
          "a.*": {"type": "integer", "multipleOf": 2}
  },
  "required": ["a","b"]
}
```

Suppose we want to generate 100 distinct objects that satisfy this schema. First, we deal with required fields (Algorithm 2 - SatReq) and we try to generate 100 distinct values for the key "a" but there are only two values possible, hence $V_a = [2, 4]$. For the next required key, which is "b" we try to generate $\lceil \frac{100}{2} \rceil = 50$ values, but the generation returns one value, $V_b = [1]$. We continue by satisfying minProperties (Algorithm 3 - SatMinProp) and for "c" we try to generate $\lceil \frac{50}{1} \rceil = 50$ but we get $V_c = [1, 2, 3, 4, 5]$, finally, we try to generate $\lceil \frac{50}{5} \rceil = 10$ for "d" by we only get $V_d = [1, 2, 3, 4]$. The number of instances we can build using the pairs key-set of values is: $M = |V_a| * |V_b| * |V_c| * |V_d| = 2 * 1 * 5 * 4 = 40$ instances.

The minProperties assertion is now met, but we are still missing $L = N - M = 60$ instances, and since the maximum number of properties per object is not reached yet which is 5, we can introduce a new property, and to this end (Algorithm 4 - MoreInstances) we pick "e", which has not yet been considered by generation. Observe that in order to obtain new records we can use values for required attributes "a" and "b" already dealt with, and combine with values for "e" either those for "c" or those for "d" or values for both "c" and "d". Each of these combinations yields a record satisfying both minProperties and maxProperties. In more details, we have $req = \{$"a", "b"$\}$, hence we set $P_{req} = |V_a| * |V_b| = 2 * 1 = 2$, and $nonReq = \{$"c", "d"$\}$, hence the subsets of $nonReq$ that we can associate to the new key "e" and to the set req of required properties is: $subs = \{\{$"c"$\}, \{$"d"$\}, \{$"c", "d"$\}\}$, hence we have: $\sum_{i=1}^{|subs|} P_i = P_1 + P_2 + P_3 = |V_c| + |V_d| + |V_c| * |V_d| = 5 + 4 + 20 = 29$. Finally, concerning the number of values for "e" we have $N_e \geq \lceil \frac{60}{2*29} \rceil$, i.e. we need two values for the new property "e" in order to reach the 100 instances, so that the generation of values for "e" returns $V_e = [1, 2]$. We perform then a cartesian product between the sets of values associated to each key in order to obtain the 100 instances.

We present now in more details the three algorithms invoked by the main object generation algorithm just illustrated.

5.2.1 Algorithm 2 - SatReq

Algorithm 2 aims at satisfying the required constraint by generating a set of key-value pairs. For each required property l, we consider all sub-schemas S_i

Algorithm 2: $SatReq(ObjTA, N)$

Result: A map from strings to sets of JSON values | $fail$

1 $objMap : Map[Str, Set[J]] = Map(); \ frac = N$
2 **for** *each* $l \in$ required **do**
3 $\quad CS = \{\|\}$
4 $\quad$ **for** *each* $(p_i, S_i) \in$ patternProperties *s.t.* $l \in L(p_i)$ **do**
5 $\quad\quad CS = CS \cup S_i$
6 $\quad \widehat{S} = merge(CS); V_l = Gen(\widehat{S}, frac)$
7 $\quad$ **if** $V_l \neq fail$ **then**
8 $\quad\quad objMap[l] = V_l; \ frac = \lceil frac / |V_l| \rceil$
9 $\quad$ **else**
10 $\quad\quad$ **return** fail
11 **return** $objMap$

---

associated to patterns p_i whose language contains l and which, therefore, need to be merged into $\widehat{S}$ using the function *merge* defined in Sub-sect. 4.2.

Going back to Example 6 we have that values generated for the key "a" must satisfy the schemas corresponding to both "a" and to "a.*", whereas values generated for the "b" key must only satisfy the schema corresponding to "b". As illustrated before, in generating those field values, we generate multiple values for each key so as to increase towards N the number of objects that we can form by combining "a" and "b" values. The pairs "a": V_a and "b": V_b, where V_a (resp. V_b) collects the set of generated values for "a" (resp. "b"), are then returned in the map $objMap$. Note that if the generation returns $fail$ for a required key, the generation for the whole schema aborts and returns $fail$.

5.2.2 Algorithm 3 - SatMinProp

The goal of Algorithm 3 is to fulfil the minProperties constraint, if not already satisfied by the previous generation step. It resorts at generating the *missing* keys from a set of candidate patterns obtained by considering the specification of propertyNames : S_{str}, if any, using $Comb(S_{str}, p_i)$ which combines the schema S_{str} with a pattern p_i expressed in patternProperties :$\{p_1 : S_1, ..., p_n : S_n\}$, resulting into a pattern that accepts strings matching both S_{str} and p_i. So we build the set P of patterns in patternProperties having non empty intersection with S_{str} (line 5) and then try to generate attributes using these patterns (lines 9–18). For any $p \in P$ we first try to generate a label l from p by means of $GenKey(p, failP)$, which returns a string value that matches p and does match any pattern in $failP$. This last one includes patterns p_i for which we cannot generate an instance for S_i. We initially set $failP$ to the empty set (line 4). Then, with $l = GenKey(p, failP)$ we recover all $p_i : S_i$ such that $l \in L(p_i)$, by including of course also $p : S$ in patternProperties (lines 4–6), and we add S_i to the sets of schemas CS, whose schemas are then merged in order to obtain $\widehat{S}$. We generate then instances V_l for $\widehat{S}$ and by definition of the merge operation

---
Algorithm 3: $SatMinProp(ObjTA, objMap, N)$

Result: A map from strings to sets of JSON values | $fail$

1   $P_{req} = \prod_{(l_i, V_i) \in objMap} |V_i|;\ m = min - |required|;\ frac = \lceil N/P_{req} \rceil$
2   **if** $m > 0$ **then**
3      $failP = \emptyset$
4      $P = \{p \mid (p, S) \in \texttt{patternProperties} \wedge \mathcal{L}(Comb(S_{str}, p)) \neq \emptyset\}$
5   **for** each $p \in P$ **do**
6   **for** $i = 1$ to m **do**
7   **if** $p \notin failP$ **then**
8                  $l = GenKey(p, failP)$
9   **if** $l == null$ **then**
10   **break**
11   **else**
12                      $CS = \{[]\}$
13   **for** each $(p_i, S_i) \in \texttt{patternProperties}$ s.t. $l \in L(p_i)$ **do**
14                          $CS = CS \cup S_i$
15                      $\widehat{S} = merge(CS),\ V_l = Gen(\widehat{S}, frac)$
16   **if** $V_l \neq fail$ **then**
17                          $objMap[l] = V_l;\ frac = \lceil frac/|V_l| \rceil;\ m = m - 1$;
18   **else**
19   **if** $|CS| == 1$ **then**
20                              $failP = failP \cup p$
21   **else**
22   **break**
23   **if** $m == 0$ **then**
24   **break**
25 **if** $m > 0$ **then**
26   **return** $fail$
27 **return** $objMap$

---

this implies that the instances in V_l satisfy all the S_i in CS (lines 14–20). So we update $objMap[l]$ and the number $frac$ of values to generate for the next key in order to reach N instances, and the remaining number m of key-value pairs to generate (lines 14–20). In case we are not able to generate any instances for $\widehat{S}$, and CS only contains the schema S (related to the pattern p of the current iteration of the loop starting at line 6, we add p to $failP$ (lines 20–21) so that we do not generate keys and related values for this pattern in subsequent iterations. After all patterns have been used to generate the missing m properties, if $m > 0$ then this means that we could not generate more properties so we return $fail$, otherwise the current mapping $objMap$ including the generated instances for a number of properties meeting `minimum` is returned.

Algorithm 4: $MoreInstances(ObjTA, objMap, N)$

Result: $M \leq N$ instances

1   $Res = \emptyset$
2   $failP = \emptyset$
3   $P = \{p \mid (p, S) \in \texttt{patternProperties} \wedge \mathcal{L}(Comb(S_{str}, p)) \neq \emptyset\}$
4   **for** each $p \in P$ **do**
5   **while** $true$ **do**
6   **if** $p \in failP$ **then**
7   **break**
8   **else**
9            $l = GenKey(p, failP)$
10   **if** $l == null \vee |Res| == N$ **then**
11   **break**
12   **else**
13               $CS = \{[]\}$
14   **for** each $(p_i, S_i) \in \texttt{patternProperties}$ s.t. $l \in L(p_i)$ **do**
15                  $CS = CS \cup S_i$
16               $(n, subs) = nbValuesToGen(ObjTA, objMap, N)$
17               $\widehat{S} = merge(CS), V_l = GenS(\widehat{S}, n)$
18   **if** $V_l \neq fail$ **then**
19                  $objMap[l] = V_l;$
20                  $Res = Res \cup getNewInstances(l, objMap, subs, N)$
21   **else**
22   **if** $|CS| == 1$ **then**
23                    $failP = failP \cup p$
24   **if** $|Res| == N$ **then**
25   **break**
26   **return** Res

5.2.3 Algorithm 4 - MoreInstances

The number of instances that can be built from the map $objMap$ once the previous phase has finished is $M = \prod_{(l_i, V_i) \in objMap} |V_i|$. These instances are constructed by taking the cartesian product of the sets V_i of values from $objMap$. The final phase of the generation, defined by Algorithm 4, is invoked when N has not been reached yet (i.e. $M < N$), and is meant to generate new instances, when possible.

Algorithm 4 is somewhat similar to Algorithm 3. The main difference is that Algorithm 4 terminates when the desired number of instances is reached or, otherwise, there are no more patterns to exploit for generating the new properties.

Like Algorithm 3, we use P and $failP$ in order to determine the pattens to use to generate new keys l. The main differences w.r.t. Algorithm 3 are the following ones. For each new generated key l we stop considering the pattern p_i from which l has been generated in case $l = null$ or in case we have a sufficient number of instances (lines 13–14). Otherwise we invoke $nbValuesToGen$ which returns the number n of values we need to generate for this key l so as to reach the desired number of instances N, and in addition it returns the set of subsets of non-required generated keys with which new values for l will be combined to in order to form new object instances (as illustrated in Example 6).

To this end $nbValuesToGen$ considers, from its input parameters, the current $M = N - \prod_{(l_i,V_i) \in objMap} |V_i|$, req and $nonReq$ as the sets of required and non-required properties appearing in $objMap$ respectively, min and max as the values of the constraints minProperties and maxProperties in the object schema $ObjTA$, and returns $subs$ as the set of subsets of $nonReq$ keys s.t. $\forall s \in subs$: $|s| \in [min - |req| - 1, max - |req| - 1]$. In other words, $subs$ is the set of all the possible combinations of non-required properties that will be associated to the new key l and to the required keys in order to generate new instances while respecting minProperties, maxProperties and required constraints. Given a new key l, the generation of N_l values for l ensures the generation of $P_{req} * N_l * \sum_{i=1}^{|subs|} P_i$ new instances, where: $P_{req} = \prod_{i=1}^{|req|} |V_i|$, where V_i is the set of values associated to the ith required key in $objMap$, and $\forall i \in [1, |subs|]$ we have $P_i = \prod_{j=1}^{|subs_i|} |V_j|$, where V_j is the set of values associated to the jth non-required key in the ith subset of $subs$ in $objMap$. To reach the N instances, we need to generate M new instances, hence, the number of values N_l to generate for the new key l should satisfy the following inequality: $P_{req} * N_l * \sum_{i=1}^{|subs|} P_i \geq M$, thus: $N_l \geq \lceil \frac{M}{P_{req} * \sum_{i=1}^{|subs|} P_i} \rceil$. If the generation produces valid JSON values for a given key l, we use $getNewInstances$ to combine the pair $l : V_l$ with the pairs that were stored in $objMap$ in order to build the new instances.

5.3 Array Types

The general form of array schemas that we consider is defined below:

$ArrTA ::=$
$\{$ type : array, minItems : $minIt$, maxItems : $maxIt$,
    prefixItems : $[S_1, ..., S_n]$, items : S,
    contains : S_c, minContains : $minC$, maxContains : $maxC$ $\}$

Dealing with arrays in the general case is slightly more involved than dealing with objects due to the upper bound assertion maxContains and to the unicity constraint uniqueItems. The upper bound assertion requires the use of negation and poses a serious limitation. Luckily enough, in our collections of schemas representing typical real-life schemas, no schema uses this assertion, so we ignore maxContainsin the current version of the work and postpone it to future work. The situation is different for the unicity constraint which is used in practice and which we can deal with by posing a restriction, that of considering the single instance generation, where the goal is to generate one instance only, i.e. $N = 1$. We will present multiple instance generation in the absence of uniqueItems then we show how we deal with uniqueItems in a restricted case where we only generate a single instance.

Remark 1. While in the generation grammar of Fig. 3, an array schema may contain a conjunction of contains assertions as consequence of applying *merge*, we decided to keep our presentation simple by restricting on the case of a single contains assertion since our solution easily generalizes to the general case where we need to take into consideration many such assertions, and because, in practice, we never encountered such a situation.

5.3.1 Multiple Array Instance Generation

The generation of N array instances is captured by Algorithm 5 which proceeds by calling $SatMinItMinC$ (cf. Algorithm 6) to generate $M \leq N$ array instances of homogenous size s satisfying both minItems and minContains, then, in case $M < N$, it attempts to generate the missing $N - M$ instances by calling $MoreInstances$ (cf. Algorithm 8) to combine the already generated M instances with extra values, if the array specification allows so, that is, if $s <$ maxItems; otherwise, it returns the instances already generated by $SatMinItMinC$.

Algorithm 5: main array generation algorithm

Data: An array assertion $ArrTA$, an integer N
Result: $M \leq N$ instances | *fail*
1 $Res = \{[]\}$
2 $arrMap = SatMinItMinC(arrTA, N)$
3 **if** $arrMap \neq fail$ **then**
4     $Res = arrProduct(arrMap);$
5     $M = |Res|;\ s = Size(Res[0]);$
6   **if** $M < N$ **and** $s <$ maxItems **then**
7   **return** $Res \cup MoreInstances(ArrTA, Res, N-M, s)$
8   **else**
9   **return** Res
10 **return** fail

Algorithm 6 - SatMinItMinC. The general idea of generating a set of N arrays from an array specification is based on generating a sequence of k fractions of N using the sub-schemas of the specification then to cross-product these k fractions in order to build the N arrays. This generation relies on preparing a candidate array CA of sub-schemas by considering the schemas of prefixItems up to *minit* and, in case prefixItems has less schemas than minItems, CA is extended with as many copies of items as needed to meet the minItems constraint (lines 4–5). This preparation also serves for preparing a potentially empty array called *tail* obtained by considering the prefixItems schemas not part of CA and which may be used for satisfying the minContains constraint (line 7).

Generation proceeds iteratively by consuming the schemas of CA in a sequential order while passing, at each iteration, the fraction $frac$ of values that remain to be generated. During this process, the contains schema is also considered towards exhausting `minContains`. In case combining this schema with the candidate schema from CA yields a valid instance, the counter j is decremented when calling $GenWithContIfPoss$ (line 9). All generated values are collected in an indexed map M and the fraction value is updated by considering the cardinality product of the already generated values, this is captured by $DomProd$ used in line 13 and defined as follows

$$DomProd(M) = \prod_{V \in values(M)} |V|$$

In case `minContains` is not completely satisfied, the generation continues by considering, this time, the $tail$ schemas, if any (line 17) until they are completely examined, then resorting to the `items` schema if its combination with `minContains` schema yields a valid value (line 19). In either cases, when generation succeeds, the generated values are appended to the map, otherwise, the whole generation fails due to failing to meet `minContains`.

Example 7. To illustrate $SatMinItMinC$, consider the following array schema and let S_1 (resp. S_2) denote the first (resp. last) sub-schema of `prefixItems`, and let S_i (resp. S_c) denote the sub-schema of `items` (resp. `contains`).

```
{ "type": "array",
  "minItems": 3,
  "prefixItems": [
          {"type": "integer", "minimum":1, "maximum":2},
          {"type" : "integer", "minimum":4, "maximum": 5}
  ],
  "items": {"type": "integer", "minimum":2, "maximum":10},
  "contains": {"type": "integer", "multipleOf":3 },
  "minContains": 2
}
```

The generation of 100 instances satisfying this schema is achieved as follows: First, CA is built as described before and yields $[S_1, S_2, S_i]$. Then generation proceeds by attempting to generate 100 distinct values from S_1 while considering S_c. However, since S_1 and S_c are incompatible, generation falls back to considering only S_1 whose maximum domain is $V_0 = \{\!|1,2|\!\}$. Now, we consider S_2 alone (after noticing that it is also incompatible with S_c) and attempt to generate $\lceil \frac{100}{|V_0|} \rceil = \lceil \frac{100}{2} \rceil = 50$ values but only succeed in producing $V_1 = \{\!|4,5|\!\}$ which contains two values. The last schema in CA is S_i which can be combined with S_c and is requested to generate $\lceil \frac{100}{|V_0|*|V_1|} \rceil = \lceil \frac{100}{4} \rceil = 25$ values while it yields $V_2 = \{\!|3,6,9|\!\}$. In order to fulfill `minContains`, we still need to generated the same set of elements $V_3 = \{\!|3,6,9|\!\}$ one more time.

Algorithm 8 - MoreInstances. The generation of missing array instances exploits the set of already generated arrays Res by extending them with new values generated by considering the array specification, to produce new instances of larger size, which are thus, distinct from those of Res. The main task of MoreInstancesis to produce $\lceil N/|Res| \rceil$ distinct values so that, when they are used for

Algorithm 6: $SatMinItMinC(ArrTA, N)$

Result: a map containing pairs (index, set of values) | $fail$

1 $M : Map[Int, Set[J]] = Map(); \; frac = N; j = minC;$
2 /* satisfy `minItems`, satisfy `minContains` if possible */
3 $m = Min(minit, size(\texttt{prefixItems})); \; CA = [S_1, .., S_m]; tail = [];$
4 **if** $size(CA) < minit$ **then**
5 $\quad\mid\;\; CA.concat(repeat(S_{it}, minit - m));$
6 **else**
7 $\quad\mid\;\; tail = [S_{m+1}, \ldots, S_n]$
8 **for** i **in** $1..size(CA)$ **do**
9 $\quad\mid\;\; (V, j) = GenWithContIfPoss(CA[i], S_c, frac, j);$
10 $\quad\mid\;\;$ **if** $V \neq fail$ **then**
11 $\quad\mid\quad\mid\;\; M[i] = V; frac = \lceil frac/DomProd(M) \rceil$
12 $\quad\mid\;\;$ **else**
13 $\quad\mid\quad\mid\;\;$ **return** fail

14 /* satisfy remaining `minContains` */
15 **while** $j > 0$ **do**
16 $\quad\mid\;\;$ **if** $size(tail) > 0$ **then**
17 $\quad\mid\quad\mid\;\; (V, j) = GenWithContIfPoss(tail.removeFirst(), S_c, frac, j);$
18 $\quad\mid\;\;$ **else**
19 $\quad\mid\quad\mid\;\; V = Gen(merge(\{|S_{it}, S_c|\}), frac); j{-}{-};$
20 $\quad\mid\;\;$ **if** $V \neq fail$ **and** $size(M) < $ `maxItems` **then**
21 $\quad\mid\quad\mid\;\; M[i] = V; frac = \lceil frac/DomProd(M) \rceil$
22 $\quad\mid\;\;$ **else**
23 $\quad\mid\quad\mid\;\;$ **return** fail

24 **return** M

---

extending the Res arrays, we are able to obtain N distinct instances. These new values are first obtained from schemas of `prefixItems` in case not all sub-schemas of `prefixItems` have been used by SatMinItMinC (lines 5–7) then by considering `items` in case all `prefixItems` sub-schemas have been used by SatMinItMinC or that the combination of the generated values does not produce N (lines 8–9). Lastly, the newly generated values are combined with the arrays from Res (lines 11 and 12), and in case the combinations exceed the requested value N, only N such combinations are returned (using an arbitrary $trunc$ function).

Example 8. continued
Now, we construct the array instances by combining elements from all V_i's whose total combination amounts to $\prod_{i=0}^{3} |V_i| = 36$; hence it remains to generate $100 - 36 = 64$ by using S_i which is requested to return $\lceil \frac{64}{\prod_{i=0}^{3} |V_i|} \rceil = 2$ values and it indeed produces $V_4 = \{2, 3\}$. These values are built and added to the previous ones, making the total number of instances generated $36 + \prod_{i=0}^{4} |V_i| = 36 + 72 = 108$ instances, and only 100 instances are returned.

Algorithm 7: $GenWithContIfPoss(S_h, S_c, N, j)$

Data: Head schema S_h, 'contains' schema S_c, # instances N, # contains j
Result: A pair (V, j), V is the generated value and updated j
1 **if** $j > 0$ **then**
2     $V = Gen(merge(\{\!|S_h, S_c|\!\}), N)$;
3   **if** $V \neq fail$ **then**
4   **return** $(V, j-1)$
5 **return** $(Gen(S_h, N), j)$

Algorithm 8: $MoreInstances(ArrTA, Res, N, m)$

Result: $M \leq N$ instances
1 $M : Map[Int, Set[J]] = Map(); CA = []; i = 0; frac = \lceil N/\,|Res|\rceil$;
2 **if** $size(\texttt{prefixItems}) > m$ **then**
3     $CA = [S_{m+1}, \ldots, S_n]$
4 /*Generate new values*/
5 **while** $i < size(CA)$ **and** $frac > 0$ **do**
6     $V = Gen(CA[i], frac); i{+}{+}$;
7   **if** $V \neq fail$ **then**
8        $M[i] = V$;
9     $frac = \lceil frac/DomProd(M) \rceil$;
10 **if** $frac > 0$ **then**
11     $V = Gen(S_{it}, frac)$;
12   **if** $V \neq fail$ **then**
13        $M[i] = V$;
14 /*combine new values with Res*/
15 $New = arrProduct(M)$;
16 **return** $trunc(N, concat(Res, New))$

5.3.2 Single Instance Generation in the Presence of uniqueItems

Generating a single array instance is a special case of multiple array instance generation, but it differs and poses more challenges because we need to ensure that the values are unique. While both processes involve satisfying minItems and minContains by invoking SatMinItMinC with $N = 1$, ensuring uniqueness of values requires handling overlapping candidate sub-schemas. This overlap necessitates trying numerous combinations of values, making the process more complex and computationally demanding. A more effective solution is to generate, for each sub-schema enough values whose combination produces a set of distinct values. This amounts finding a perfect matching in a bipartite graph mapping indexes to values. We adopt this solution and use the well-known Hall condition [12] for deciding the existence of a perfect matching while dynamically building the bipartite graph by considering the schemas from prefixItems and items.

Example 9. To illustrate the process of building the bipartite graph, consider the following schema which features all kinds of assertions that can be expressed in arrays while requiring the values to be distinct.

```
{ "type": "array", "minItems": 2, "maxItems":4, "minContains": 2,
  "prefixItems": [ {"type": "integer", "maximum":2}, {"enum" : [3,12,18] } ],
  "items": {"type": "integer", "multipleOf":2, "maximum": 20},
  "contains": {"minimum": 10, "multipleOf":3 },
  "uniqueItems": true
}
```

To satisfy this schema, an array must have a minimum of two items, the first of which adhere to prefixItems and it must contain at least two items satisfying the contains schema, which in our case can only be satisfied by generating an addition item using items since the first schema of prefixItems contradicts that of contains.

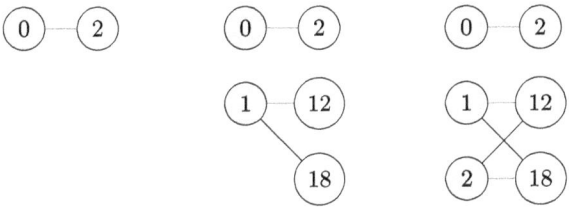

Fig. 5. Graph G at iterations: 0, 1 and 2

Figure 5 depicts the steps for building the graph: initially, only the first schema of prefixItems is considered and generates the set $\{2\}$, then the second schema is considered together with contains and leads to generate $\{12, 18\}$, then, items is considered and combined with contains to fulfil minContains and leads to generating exactly the same set $\{12, 18\}$. The graph admits two perfect matchings: $\{0 \to 2, 1 \to 12, 2 \to 18\}$ and $\{0 \to 2, 1 \to 18, 2 \to 12\}$ and allows for generating two arrays of unique items: $[2, 12, 18]$ and $[2, 18, 12]$.

Algorithm 9 describes the approach for constructing the graph G, represented as the map M. As stated previously, the process of generating one array instance in the presence of uniqueItems is similar to generating multiple arrays and only differs in the number of values to generate for each candidate of CA and in the verification of the existence of the perfect matching. It is important to note that we omitted the instructions for dealing with contains and minContains from the algorithm, as their satisfaction is handled similarly to the case of generating multiple instances.

During this process, the Hall condition is checked if the required number of values to generate has not been attained; finally, a greedy approach is used to find the perfect matching from the a sub-graph of the bipartite graph represented by the map M.

Algorithm 9: UniqueItems generation

Data: An array of schema CA
Result: An instance $J_a \mid fail$

1 $M : Map[int, Set[J]] = Map(); \ i = 0$
2 **while** $i < |CA|$ **do**
3 $\quad V_i = GenS(CA[i], i+1)$
4 $\quad$ **if** $V_i == fail$ **then**
5 $\quad\quad$ **return** $fail$
6 $\quad$ **else**
7 $\quad\quad$ **if** $|V_i| < i+1$ **then**
8 $\quad\quad\quad$ **if** $\neg existsPerfectMatch(M)$ **then**
9 $\quad\quad\quad\quad$ **return** $fail$
10 $\quad\quad M[i] = V_i; \ i = i+1$
11 **return** $PerfectMatch(M)$

6 Experiments

The goal of the experiments is manifold: i) to compare our approach with existing data generation tools in terms of the generated instances *correctness* and in terms of execution times, ii) to measure the effectiveness of our approach for generating sets of valid instances of varying cardinality size, and finally, iii) to assess the ability of our approach to deal with uniqueItems which is an original contribution of our work. To achieve these goals, we implemented our approach in Scala[1] exploiting pattern matching for encoding the schema rewriting rules used for pre-processing the input schema and by adopting a more imperative style for implementing the generation algorithms. We fully relied on the Brics library [13] for generating strings from patterns following the same design choices of our previous work on *witness generation* [8]. We carried out the experiments on a Precision 7550 laptop with a 12-core Intel i7 2.70GHz CPU, 32 GB of RAM, running Ubuntu 23.04. We used both real-world and synthetic schema collections that we describe below.

6.1 Comparative Experiments

To compare our approach with the tools described in Sect. 2, we use the schema collections from [8] which proved useful for testing the effectiveness and efficiency of *witness generation* approach defined there. Three of these collections consist of real-world schemas which were obtained from public sources like Github and one collection has been synthetically generated from the JSON Schemaofficial Test Suite [4] in order to test schema containment. The real-world schemas collections are diverse enough and help in assessing the coverage of our approach in terms

---

[1] The source code is available at https://gitlab.lip6.fr/attouche/jschemadatagenerator.

of JSON Schemafeatures. In turn, the synthetic schemas make extensive use of negation and allow for evaluating the limitations of our approach in terms of support for negation. We report about the general statistics of these collections in Table 2 and distinguish between satisfiable and unsatisfiable schemas, when relevant.

Since our study is meant to compare w.r.t. tools that generate *a single* instance while ignoring `uniqueItems`, we limit ourselves, in this experiment, to the generation of *one* instance by setting $N = 1$, while annealing the effect of `uniqueItems` in the original schemas, by rewriting each of its occurrences into a non-validating keyword.

Table 2. Description of the schema collections [8]

Collection	#Total	#Sat/#Unsat	Size (KB): Avg/Max
Github (Git)	6,413	6,373/40	8.7/1,145
Kubernetes (K8s)	1,092	1,087/5	24.0/1,310.7
Snowplow (Snw)	420	420/0	3.8/54.8
WashingtonPost (WP)	125	125/0	21.1/141.7
Containment-draft4 (CC)	1,331	450/881	0.5/2.9

6.1.1 Correctness Analysis

To evaluate the ability of our approach for generating valid instances, we run our tool and the three other tools on each schema, then collected the generated instances and validated them against the original (satisfiable) schema using an external validator [5]. We measured the correctness ratio for each combination of tool and schema collection and report the results in Table 3. More precisely, we indicate the ratios for: schemas that are processed and yield a valid witness (*Success*), schemas that are processed by yielding a non-valid witness (*Logical Error*), and those that encounter a runtime error during processing (*RT Error*). We also report about the ratio of schemas that our tool intentionally discards to avoid yielding potentially invalid instances (*Failure*). To ensure the termination of our experiments we set a timeout of 300 s for processing each schema.

The results show many interesting trends: our tool outperforms all the others for almost all collections in terms of correctness and features a limited portion of *Failures*. The case of Snowplow is one exception for which *JF* has a slightly better performance than our tool which has more logical errors. Our investigation leads us to conclude that all our logical errors are related to mishandling of patterns by the Brics library which fails in the presence of complex patterns. On the other side, our tool shows an important improvement over the others in *containment* collection which makes intensive use of negation for which we have a partial handling while other tools tend to ignore this operator.

Table 3. Correctness results and execution times.

Coll.	Tool	Success	RT Error (+Failure)	Logical Error	Time (ms) med./avg.
Git	Ours	96.72%	0.89% +0.75%	1.64%	1/176
	DG	94.21%	2.86%	2.93%	22/228
	JF	82.64%	5.80%	11.56%	117/323
	JE	58.68%	20.77%	20.55%	9/10
K8s	Ours	100%	0% +0%	0%	1/23
	DG	99.54%	0%	0.46%	22/31
	JF	89.84%	0.18%	9.98%	122/125
	JE	69.41%	8.43%	22.16%	7/8
Snw	Ours	94.76%	0.48% +0.48%	4.28%	2/226
	DG	94.76%	0%	5.24%	24/30
	JF	95.24%	2.62%	2.14%	119/549
	JE	72.38%	18.10%	9.52%	2/5
WP	Ours	100%	0% +0%	0%	5/96
	DG	96.80%	0%	3.20%	29/42
	JF	87.20%	0%	12.80%	121/134
	JE	29.60%	25.60%	44.80%	6/11
CC	Ours	77.76%	5.03% +14.05%	3.16%	1/45
	DG	28.77%	30.88%	40.35%	20/22
	JF	27.20%	2.78%	70.02%	206/219
	JE	0.23%	3.91%	95.86%	1/508

Finally, we examined the *Failures* cases and confirmed that all discarded schemas feature a negation that can be eliminated, like negation of object and arrays.

6.1.2 Execution Time

We report about running time in the same table by indicating the median and average values, for each combination of collection and tool. We observe that the tools behave differently on different collections, which makes it hard to generalize about the total speed of the tools. For a fair comparison, we need to relate these numbers with the success rate of each tool, since, failure to process a schema may take a fewer time than processing the same schema. We also would like to stress that the sound-and-complete technique in [8] suffers from higher execution times and requires a total of 3 h to process the five collections whereas our optimistic tool takes around 20 min.

Since we are interested in improving our tool from the perspective of efficiency, as well, we investigated the reasons which lead to a higher runtime in Snowplow and discovered a rather intensive use of complex patterns that our implementation delegates to the Brics [13] library and misses, thus, an optimization opportunity.

6.2 Multiple Instance Generation Experiments

To evaluate the ability of our approach to generate N distinct instances, we use the largest real-world collection from the previous experiment, and vary N from 10 to 1000 using a step of 10, up to 100, then increasing this step to 50 until 250, and finally doubling the number N from 250 to 1000. For each value of N, we discard from the input schemas those that are not able to generate N instances by design, and that we detect by examining the logs upon running our generation and by inspecting the schemas manually.

We run the generation on all schemas, for each value of N by setting a timeout of 300 s per schema. We measure the success rate by considering the ratio of schemas whose generation succeeds and whose generated values are all valid. We also measure runtime errors, failures and logical errors and report all these measurements in Table 4.

We make the following observations: (i) the success ratio remains rather stable with a minor degradation on higher values of N in favour of a slight increase in runtime errors and logical errors, and (ii) the failure ratio is rather stable and around 1%. By examining the logs, we discovered that the main reasons of runtime errors are related to the timeout being exceeded while most logical errors are due to the generation of non valid strings from patterns using special characters that the Brics library does not natively support. By investigating the failures we realized that the failing schemas feature negation either explicitly using `not` or implicitly using `oneOf` and these negations appear on *required* branches of the schema which could not be avoided by our generation strategy. The other failures are related to some limitations of the Brics library which was not able, for some patterns, to generate the required number of (distinct) strings despite the fact that the language of the input pattern satisfies the cardinality constraint.

The reported execution time confirm the efficiency of the generation which is able to produce 1000 instance in 3 s, on average. The measures about the file size show a linear increase of the size w.r.t. N which is expected since our generation is deterministic by nature.

6.3 `uniqueItems` Experiments

To assess the ability of our approach for dealing with `uniqueItems`, we built a dedicated collection of synthetic schemas that describe arrays where `uniqueItems` is always set to true. Each schema has a `prefixItems` consisting of five sub-schemas all referring to the same type (number, string, object or array) but which are mutually different since they associated with varying constraints pertaining to their types (for instance, for array of numbers, we vary the bounds of the number specification, while for arrays of strings, we vary the patterns). In addition to `prefixItems`, each schema has non-empty `items` and `contains` assertions whose sub-schemas are compatible with those of `prefixItems`. For each family of schemas, we use bounds with varying values: `minItems` ranges from 5 to 10, while `minContains` ranges from 2 to 8.

Table 4. Multiple instance generation results

N	#Schemas	Success	RT Error	Failure	Logical Error	Time (ms) Avg/Med	Avg Size (KB)
10	6,224	93.83%	0.64%	1.03%	4.50%	1,598/1	1.35
20	6,215	93.63%	0.66%	1.05%	4.66%	1,621/2	2.74
30	6,215	93.58%	0.67%	1.05%	4.70%	1,654/2	4.15
40	6,215	92.85%	1.48%	1.05%	4.62%	1,675/3	5.54
50	6,215	92.85%	1.48%	1.05%	4.62%	1,676/4	6.94
60	6,215	92.82%	1.48%	1.05%	4.65%	1,682/5	8.34
70	6,214	92.77%	1.48%	1.05%	4.70%	1,762/9	9.76
80	6,214	92.71%	1.53%	1.05%	4.71%	1,772/12	11.17
90	6,212	92.71%	1.53%	1.05%	4.71%	1,776/14	12.58
100	6,212	92.66%	1.58%	1.05%	4.71%	1,793/17	14
150	6,210	92.58%	1.64%	1.05%	4.73%	1,918 /23	21.12
250	6,207	92.56%	1.69%	1.01%	4.74%	2,004 /61	35.44
500	6,203	92.55%	1.71%	1.02%	4.72%	2,277/244	71.72
1000	6,201	92.52%	1.76%	1.01%	4.71%	3,024 /949	145.98

The collection comprises 35,366 schemas in total distributed as follows, by considering the content of their array-schema: 5,962 with arrays of numbers (ANum), 5,610 with arrays of strings (AStr), 7,688 with arrays of objects (AObj) and 16,106 with arrays of arrays (AArr). The average size of instances generated from schemas for each sub-collection are reported in Table 5 together with the standard metrics about correctness and execution time.

Table 5. uniqueItems experiments

Coll.(size)	Success	RT Error	Failure	Logical Error	Avg time (ms)	Avg Size (KB)
ANum(5,962)	100%	0%	0%	0%	1.51	0.04
AStr(5,610)	100%	0%	0%	0%	5	0.12
AObj(7,688)	100%	0%	0%	0%	7.64	0.47
AArr(16,106)	100%	0%	0%	0%	13.07	0.38

We observe that our generation succeeds for all schemas in the four sub-collections and that it is very fast at processing them (highest average time is 13 m sec). We also observe that execution time increases with the size of the generated instances, which is expected with the exception of arrays nested in arrays since the arrays of our collection feature a non-empty `contains` with an argument of `minContains` exceeding 2, which requires to combine, during generation, a sub-schema of `contains` with the candidate list built from `prefixItems`.

7 Conclusion

In this paper we developed an *optimistic* data generation approach for JSON Schema. Our goal was to find a compromise between efficiency and completeness, since the theoretical complexity of the problem is prohibitive. To achieve our goal, we depart from the sound and complete solution developed in previous work [8] which relies on expensive schema preparation. Instead, we adopt a simpler schema rewriting technique which exhibits the parts of the input schema that are more amenable to generation, then we proceed by generate instance, immediately, while striving to satisfy all constraints during the generation itself. Extensive experiments showed that, our technique is efficient, and covers a large set of real-life schemas while it remains efficient on synthetic schemas.

In the future, we plan to adapt our technique to allow for generating more realistic instances by taking into account data distribution and user preferences. We also plan to improve our implementation to address the limitations concerning string generation for complex patterns of the underlying BRICS library. This can be achieved by translating JSON Schema regular expressions into BRICS-compatible regular expressions, thereby enhancing the ability to generate strings that match complex patterns accurately.

References

1. Json everything. https://github.com/gregsdennis/json-everything
2. Json faker. https://github.com/json-schema-faker/json-schema-faker
3. Json generator. https://github.com/jimblackler/jsongenerator
4. Json schema test suite (2020). https://github.com/json-schema-org/JSON-Schema-Test-Suite/blob/master/tests/draft6/additionalProperties.json
5. Json schema validator (2022). Accessed 19 Sept 2022
6. Attouche, L., et al.: A tool for json schema witness generation (2021)
7. Attouche, L., Baazizi, M.A., Colazzo, D.: Overview and perspectives for optimistic JSON schema witness generarion. In: 23èmes Journées Bases de Données Avancées, BDA (2023)
8. Attouche, L., Baazizi, M.A., Colazzo, D., Ghelli, G., Sartiani, C., Scherzinger, S.: Witness generation for JSON schema. Proc. VLDB Endow. **15**(13), 4002–4014 (2022)
9. Baazizi, M.A., Colazzo, D., Ghelli, G., Sartiani, C., Scherzinger, S.: Negation-closure for JSON schema. Theor. Comput. Sci. **955**, 113823 (2023)
10. Bourhis, P., Reutter, J.L., Suárez, F., Vrgoc, D.: JSON: data model, query languages and schema specification. In: Sallinger, E., Van den Bussche, J., Geerts, F. (eds.) PODS, pp. 123–135. ACM (2017)
11. Habib, A., Shinnar, A., Hirzel, M., Pradel, M.: Finding data compatibility bugs with JSON subschema checking. In: ISSTA 2021: 30th ACM SIGSOFT International Symposium on Software Testing and Analysis, Virtual Event, Denmark, 11–17 July 2021, pp. 620–632 (2021)

12. Hall, P.: On representatives of subsets. J. London Math. Soc. 26–30 (1935)
13. Møller, A.: dk.brics.automaton – Finite-State Automata and Regular Expressions for Java (2021). https://www.brics.dk/automaton/. Accessed 19 Sept 2022
14. Wright, A., Andrews, H., Hutton, B.: JSON schema validation: a vocabulary for structural validation of json - draft-handrews-json-schema-validation-02. Technical report, Internet Engineering Task Force (2019)

Author Index

A
Amann, Bernd 76
Amer-Yahia, Sihem 1
Attouche, Lyes 119

B
Baazizi, Mohamed-Amine 119
Benouaret, Idir 1
Bouarour, Nassim 1
Bouganim, Luc 37

C
Colazzo, Dario 119
Constantin, Camelia 76

M
Mirval, Julien 37

N
Naacke, Hubert 76

O
Ogasawara, Eduardo 98

P
Porto, Fábio 98

R
Rahimi, Hamed 76

S
Sandu Popa, Iulian 37

V
Valduriez, Patrick 98

Z
Zorrilla, Rocío 98

GPSR Compliance
The European Union's (EU) General Product Safety Regulation (GPSR) is a set of rules that requires consumer products to be safe and our obligations to ensure this.

If you have any concerns about our products, you can contact us on

ProductSafety@springernature.com

In case Publisher is established outside the EU, the EU authorized representative is:

Springer Nature Customer Service Center GmbH
Europaplatz 3
69115 Heidelberg, Germany

www.ingramcontent.com/pod-product-compliance
Ingram Content Group UK Ltd.
Pitfield, Milton Keynes, MK11 3LW, UK
UKHW022241230426
12048UKWH00018BA/1398